Geology Field Guide to Northern California

# K/H GEOLOGY FIELD GUIDE SERIES

# FIELD GUIDE
# Northern California

### John W. Harbaugh
Stanford University
Stanford, California

**KENDALL/HUNT PUBLISHING COMPANY**
Dubuque, Iowa

K/H GEOLOGY FIELD GUIDE SERIES

   Consulting Editor
   *John W. Harbaugh*
   *Stanford University*

Copyright © 1974 by Wm. C. Brown Company Publishers

Copyright © 1975 by Kendall/Hunt Publishing Company

Library of Congress Catalog Card Number: 73—88485

ISBN 0—8403—1273—3

All rights reserved. No part of this publication may be reproduced, stored in a retrieval system, or transmitted, in any form or by any means, electronic, mechanical, photocopying, recording, or otherwise, without the prior written permission of the copyright owner.

Printed in the United States of America

# Contents

*Preface* — vii

*Introduction* — xi

*Chapter 1*  Coast Ranges Province: Point Reyes Field Trip — 1

*Chapter 2*  Coast Ranges: Santa Cruz Mountains Coastal Area Field Trip — 18

*Chapter 3*  Sierra Nevada and Great Basin Provinces: Yosemite-Mono Craters Field Trip — 43

*Chapter 4*  Cascade, Modoc Plateau, and Klamath Provinces: Lassen Shasta Field Trip — 84

*Appendix A*  The Geological Time Scale — 119

*Index* — 121

# Preface

Dr. Harbaugh's book is one of a series of books in the K/H *Geology Field Guide Series*. The purpose of this series is to provide a layman's guide to outstanding geologic features of each region treated. The first book in the series has been written by Professor Robert P. Sharp of the California Institute of Technology. His book provides a guide to some of the geologic features of southern California. My book is complementary in that it treats northern California. While both books provide a general overview of their respective regions, they are highly selective in the particular geologic features that they treat and in the specific field trip guides that they include.

I have organized the book according to the geologic provinces of northern California, briefly treating the Coast Ranges, Sierra Nevada, Cascade, Modoc Plateau, Klamath, and Great Basin provinces. My approach has been to outline the general characteristics of each province, and to describe a series of field trip routes and stops which are organized by geologic provinces. My main purpose, however, is to provide the layman with a guide to some of the localities of outstanding geological interest in northern California, and to discuss the relevance of these localities to the region. As a consequence, I have intermingled comments and discussion on geologic features, with logistical details of getting from one field trip stop to another.

While the entire northern half of California is broadly treated in this book, there is no attempt to provide systematic geographic coverage over the region as a whole. Northern California is much too large and too geologically complex to permit uniform coverage in a book of this size. Instead, I have chosen to concentrate on selected areas that contain examples of geological features which may be conveniently observed with a coordinated sequence of routes and stops on field trips. For reasons of space, I have omitted consideration of many localities that would otherwise be suitable for additional field trips, as for example, the Santa Lucia Mountains, the Big Sur coastal features and the recent volcanic features around Clear Lake.

This book assumes that the reader has (or is acquiring) an elementary knowledge of geology. An introductory course in geology should provide enough background for most readers to make effective use of the book. The book should also be usable by those persons who have never formally studied geology, if they are willing to spend some effort studying an introductory text. There is no better way to stimulate an interest in geology than to travel, observe, and interpret the geologic features seen enroute. The real measure of the success of this book, therefore, is the degree to which it stimulates the reader to see and interpret for himself.

I wish to emphasize that this book has been written for beginning geology students and for amateur geologists—not for professional geologists, although they may find it useful. This conscious effort to reach the beginning geologist has strongly influenced the selection of geologic features that are described in the book and the manner in which they are described, I have tried to emphasize those features that can be more or less readily understood and appreciated by beginners. Every professional geologist knows, however, that interpretation of many geologic features depends on evidence that is relatively subtle. For example, the preparation of a geologic map in an area of poor exposures may involve piecing together scraps of information from scattered outcrops. While the map may portray very significant geologic features, such as a major fault or a large fold, these features may be of little immediate interest to beginning geology students because they are difficult to visualize. My philosophy in this book, then, is to attempt to guide the reader to those localities where the geologic features are relatively well displayed. Furthermore, in places I have dwelled on rather common, "garden variety" geologic features that may be of scant importance globally speaking, but which should be interesting for beginners. Finally, discussions of some of the important theories that are currently being debated, such as evolution of the continents, continental drift, and plate tectonics have beeen deliberately avoided. I am fully aware that California geology has important evidence that bears strongly on these emberging theories. By avoiding these topics, however, I do not imply that they are not relevant to the study of California geology; instead I regard them as beyond the scope of this book.

## ACKNOWLEDGMENTS

I am indebted to various individuals for assistance. The manuscript was typed by Miss Pat Hishiki. Perfecto Mary made most of the drawings and printed virtually all of the photographs. A number of the photographs were taken by the late Eliot Blackwelder and were selected from his collection which was left to Stanford University. The manuscript was read by Gordon B. Oakeshott, Benjamin M. Page, and Carl Hauge. Richard Harbaugh helped with the field observations. To all of them, my thanks.

# Introduction

A quick inspection of California's relief map (Map 1) is enough to suggest that California can be divided into areas or regions of contrasting topography. The Sierra Nevada region, for example, contrasts strongly with the Great Valley. These differences in topography reflect differences in the underlying geology. For example, the types of rocks in the Great Valley and their structural configuration are quite different from those of the Sierra Nevada. It is logical, therefore, to distinguish the differing regions on the basis of their general geological characteristics. These geologic regions are termed geologic provinces. Eleven main geological provinces have been distinguished in California. From northwest to southeast, they are: the Klamath Mountains, Cascade Range, Modoc Plateau, Great Basin, Coast Ranges, Great Valley, Sierra Nevada, Mojave Desert, Transverse Ranges, Peninsular Ranges, and the Salton Trough. These provinces are somewhat arbitrarily defined, and they can be subdivided into lesser provinces. On the other hand, each is broadly distinguishable from the others in terms of topography, large-scale geologic structure, and to some extent, by type and age of rocks. It would be difficult, however, to frame a precise definition or description of each province, for all the provinces have many major types of geologic features in common. We can regard the provinces as defined partly as a matter of simple convenience. The geology within each province can be shown on geologic maps in varying levels of detail. A much simplified geologic map of California is shown in Map 2. A more detailed, colored geologic map that will be useful to readers of this book is available for purchase from the U.S. Geological Survey. It is U.S. Geological Survey Miscellaneous Map Investigations Map I-512, which may be purchased for 25 cents by an order directed to U.S. Geological Survey, Distribution Sec-

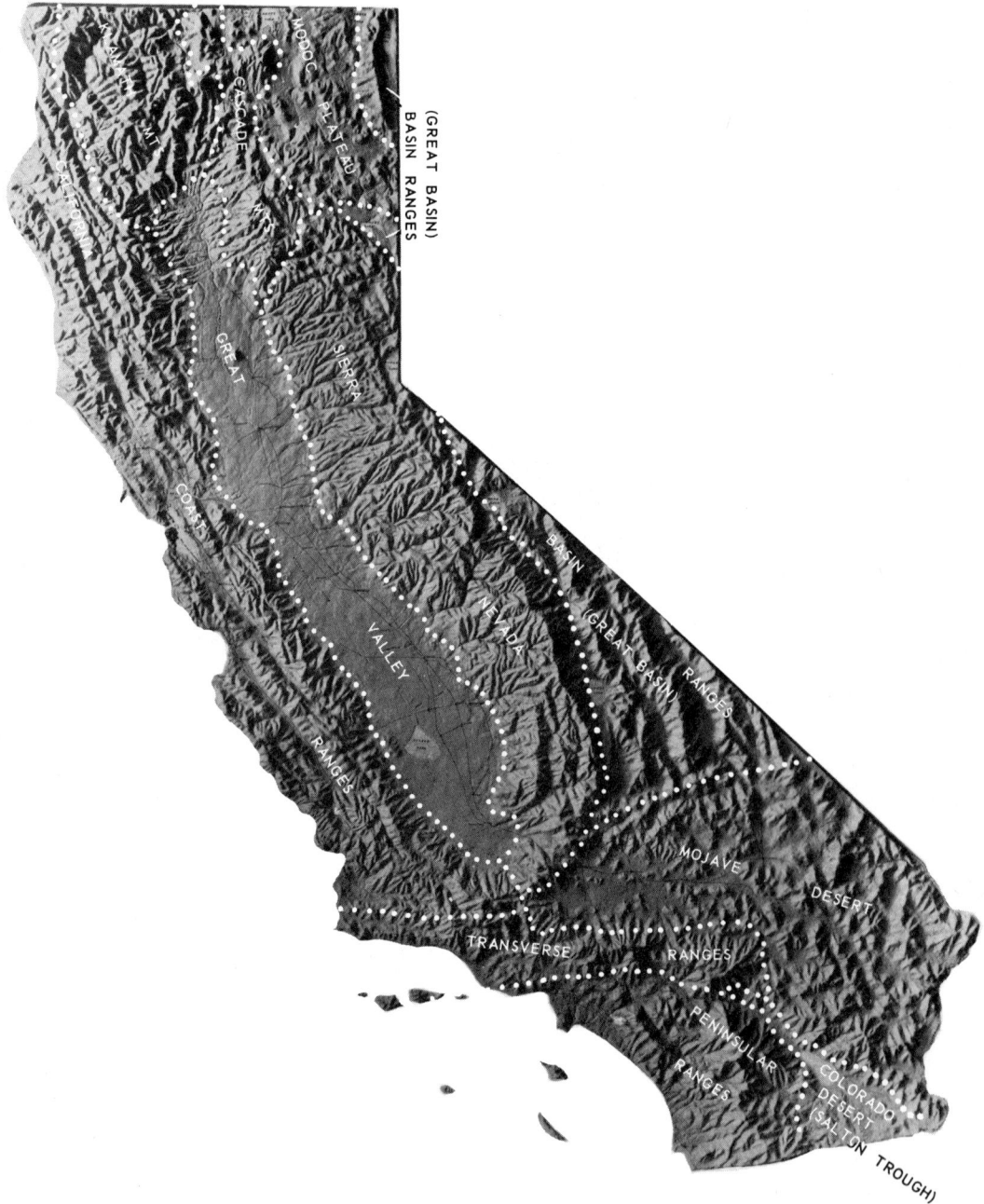

**Map 1.** Relief map of California. (From California Division of Mines and Geology.)

THE GEOLOGIC PROVINCES • xiii

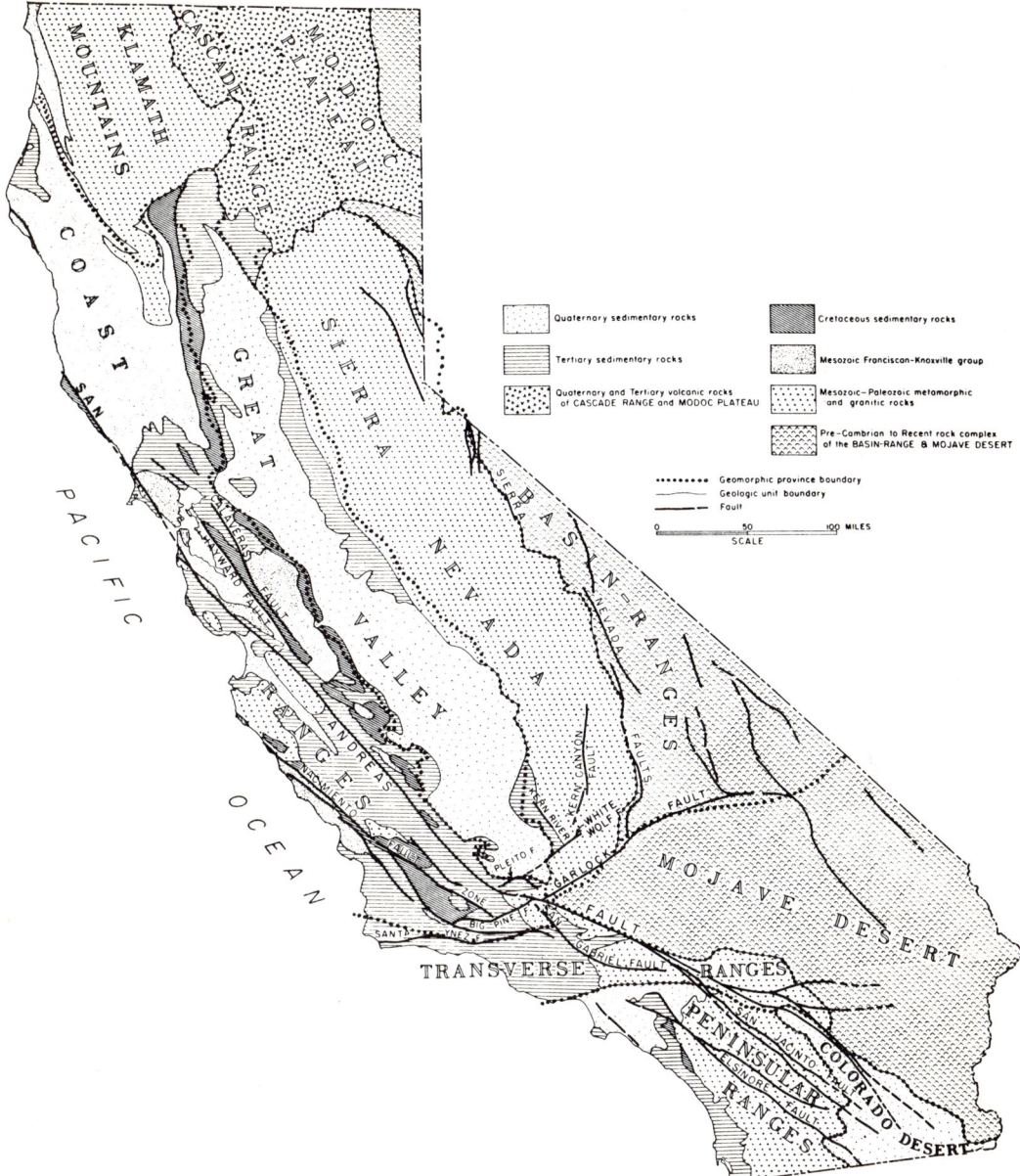

**Map 2.** Greatly simplified geologic map of California showing major groups of rocks. Boundaries of geologic provinces are shown with dotted lines. (From California Division of Mines and Geology.)

xiv • INTRODUCTION

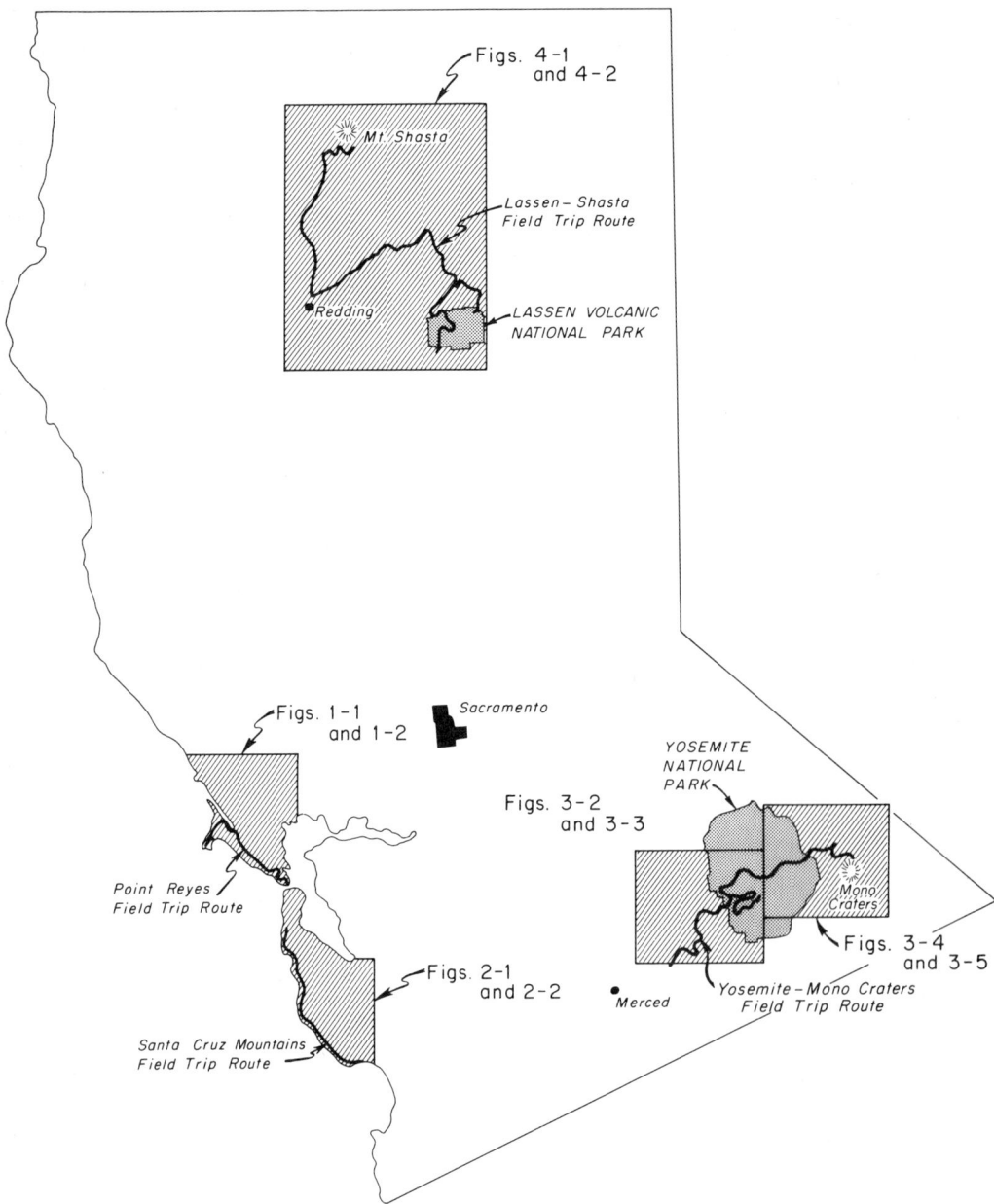

**Map 3.** Map outlining routes of field trips described in this book. Map also shows outlines of areas for which geologic maps of intermediate detail are presented in this book.

tion, Federal Center, Denver, Colorado 80255, or may be ordered from California Division of Mines and Geology, P.O. Box 2980, Sacramento, California 95812.

Map 3 outlines the routes of field trips in northern California that are described in this book. There is no prescribed sequence in which the various field trips can be taken. In a general way, their sequence in this book is related to increasing distance away from the San Francisco Bay area, but this is an artifice of convenience.

Chapter 1

# Coast Ranges Province: Point Reyes Field Trip

## COAST RANGES PROVINCE

The Coast Ranges province occupies a greater proportion of the area of California than any other province (Map 1). The Coast Ranges trend north-northwesterly, and form the coastal border along nearly three-fourths the length of California's coast line. The Coast Ranges province grades into the Transverse Ranges at the south, and adjoins the Great Valley on the east over much of its expanse. In extreme northwestern California, the Coast Ranges province narrows to about 10 miles where it lies just west of the Klamath Mountains province, and continues into Oregon. The boundaries between these provinces are not sharply defined, such as political boundaries between states. Thus, the sharp lines on Map 1 must be regarded as delineating the generalized boundaries between provinces.

The Coast Ranges province consists of a complex of lesser, more or less independent mountain ranges and valleys, many of which have specific names. Some of the ranges extend continuously for a hundred miles or more, as for example, the Mendocino Range in the northern part of the Coast Ranges, and the Diablo Range, which with the Santa Cruz Range bound the topographic lowland containing part of San Francisco Bay. Other mountain ranges include the Gabilan Range and the Santa Lucia Range, which flank the two sides of the Salinas Valley. All of these ranges and valleys, including many not specifically mentioned above, are characterized by exceedingly complex geologic structure. In turn, the structural complications influence their topography, although their intricate topography reflects a variety of geologic influences in addition to those of their internal structure. An overview of the geology of the Coast Ranges is provided by Page (pp. 253-75) in the symposium volume, *Geology of Northern California* (Bailey, ed. 1966), which also contains a number of other articles that deal with the Coast Ranges.

The topographic relief in the Coast Ranges is appreciable, although it falls short of the total relief in parts of the Sierra Nevada province, for example. The highest peaks in the Coast Ranges are seldom more than 6,000 feet above sea level, but considering that some peaks rise almost directly from sea level, the relative relief and the steep slopes locally provide some of the most rugged topography in California. Much of the Coast Ranges province, however, is characterized by topography that is much less rugged than is typical of high, glaciated mountains, for example. More commonly the Coast Ranges mountains are rounded, and have, shall we say, a "mellow look."

The "grain" of the Coast Ranges is clearly portrayed in relief Map 1. The individual ranges and valleys generally extend in a northwest-southeasterly direction. The coastline itself is irregular, but trends somewhat more northerly than the individual ranges. As a consequence, many of the geologic structures represented by the ranges and valleys extend beyond the shoreline, out into the submerged continental shelf and continental slope, which forms the true edge of the continent. If we define the western boundary of California as lying offshore, we can add two more geologic provinces to our list, namely the continental shelf and the continental slope. In certain respects, however, these offshore provinces could be considered extensions of the Coast Ranges.

In northern California, the continental shelf varies greatly in width. It extends westward from the shoreline to depths of about 600 feet and it slopes about three degrees. Beyond that, the seafloor is generally steeper, and the continental shelf passes into the continental slope, which extends downward to depths of about 10,000 feet, where the slope flattens out to form the deep-sea plain.

The directions of streams of the Coast Ranges generally follow the grain of the region, tending to flow in a northwesterly or southeasterly direction. By comparison with the streams that flow into the Great Valley from the Sierra Nevada, most of the streams of the Coast Ranges are smaller. The Eel River, which drains the Mendocino Range, is the largest in terms of volume of water. Other important streams include the Russian River and the Salinas River.

## GEOLOGIC CORE COMPLEXES OF THE COAST RANGES

The Coast Ranges province is characterized by two entirely different kinds of core complexes. A core complex is regarded as the aggregation of old rocks that forms the geologic "basement." In a given region the core complex is the oldest group of rocks which we can observe directly. The rocks of the core complexes may be exposed at the surface, or in many places, they lie beneath younger rocks. We have no way of observing directly what lies beneath the core complexes. One of the core complexes consists of the Franciscan assemblage which is an extremely complicated mass of various kinds of sedimentary and altered volcanic rock deposited during the Jurassic and Cretaceous Periods. (The geologic periods and their subdivisions, which form the geologic time scale are listed in the appendix.) The other consists of metamorphosed sedimentary rocks that have been intruded by masses of granite, and other kinds of igneous rock that are related to granite in composition. One of the most interesting and puzzling aspects about these two complexes is that they never (insofar as we know) occur together except where brought alongside one another by fault movements. In other words, where the granitic-metamorphic core complex occurs, the Franciscan assemblage is generally absent, and vice versa. This mutual exclusiveness of the two basement complexes has deep significance and is closely related to mountain building, earthquakes, and continental drift.

The core complexes are overlain by younger rocks ranging from Upper Cretaceous to Quaternary in age. Their outcrops are shown in Map 2. Details of some of the overlying rocks are discussed in the individual trip stop descriptions that follow.

**Granitic-Metamorphic Core Complex**

The basement or core complex composed of granitic and metamorphic rocks is exposed in many places in the Coast Ranges, outcropping intermittently from the juncture of the Coast Ranges with the Transverse Ranges northwestward as far as Bodega Head, about 20 miles north of Point Reyes Peninsula. The areal distribution of the basement rocks is shown in the generalized geologic Map 2.

Of the two principal types of rocks in the granitic-metamorphic core complex, the metamorphic rocks are unquestionably the older. However, the age or ages of the metamorphic rocks are known only in a relative, rather than an absolute sense. They are older than the granitic rocks that intrude them, and the ages of the granitic rocks are provisionally in the range of about 80 to 110 million years. It is possible that the metamorphic rocks are only slightly older than the igneous rocks, but this is only guarded speculation. It is also possible that they are much older, perhaps having been deposited during the Paleozoic Era. Poorly preserved fossils found in the metamorphic rocks suggest a Paleozoic age. Some of the best exposures of metamorphic rocks in the Coast Ranges province occur in the Santa Lucia Range. Here the name Sur Series has been applied (the name being derived from the Big Sur and Point Sur localities), but the Sur Series has been expanded to include all metamorphic rocks elsewhere in the Coast Ranges. The rocks of the Sur Series include gneiss, schist, quartzite, marble, and in places, only slightly metamorphosed siltstone.

The metamorphic rocks range widely in metamorphic grade. In the Santa Cruz Range, for example, relatively weakly metamorphosed siltstones are present, whereas in the Santa Lucia Range the rocks are of higher metamorphic grade, as for example, in the Arroyo Seco area west of Greenfield. The original sedimentary rocks that were subsequently metamorphosed undoubtedly were common types of rocks, including sandstone (now quartzite), limestone (now marble), shale and siltstone (now schist). In many places, layering is readily apparent in the metamorphic rocks. Some of the layering is a relic of the original sedimentary bedding, whereas other layering has been produced in the metamorphic processes.

The granitic rocks of the Coast Ranges crop out over an extensive area. In general, the granitic rocks form large masses or plutons that intrude the metamorphic rocks. Some of the granitic masses are large, extending over several hundred square miles or more. Others are smaller, but in many examples, their areal extent is not known because they are buried beneath younger rocks and cannot be fully observed. It is probable that the granitic-metamorphic complex exists beneath all of the Coast Ranges province where the Franciscan assemblage does not occur. Consequently, we would expect to encounter granitic or metamorphic rocks if we drill through the Cenozoic and older rocks in the Coast Ranges in those areas where the Franciscan assemblage does not occur.

The granitic bodies are not simple, homogeneous masses of granite. Close scrutiny indicates that some of the large masses of granite consist of an aggregation of multiple plutons that differ slightly in age with respect to each other. The plutons include a variety of rocks, including potash-rich granite, as well as quartz diorite and granodiorite. In places, smaller igneous bodies are present which consist of gabbro and other dark igneous rocks that contrast strongly in composition and appearance with the granitic and granite-like rocks.

The granitic rocks have been dated by radiometric methods and have yielded potassium-argon dates in the 80 to 90 million-year range. The dates may be reliable, but some geologists have voiced suspicion that these ages are minimum ages, that is, that the rocks cannot be younger than about 80 or 90 million years old, but could be substantially older. The potassium-argon method of radiometric dating depends on the retention of argon that has been produced by decay of radioactive potassium within crystals in the rock. If argon produced by radioactive decay has leaked away, age determinations tend to be erroneously low. Unfortunately, the tendency for argon to leak away is greater at higher temperatures, and it is possible that the calculated ages may represent the last interval of heating instead of actual intrusion. However, some dates by the rubidium-strontium method tend to confirm an upper age limit of about 100 to 110 million years.

## Franciscan Core Complex

The Franciscan complex forms the core complex elsewhere in the Coast Ranges. It is true that the Franciscan complex locally occurs in close proximity to the granitic-metamorphic complex. But, where this is the case, the two complexes are invariably in fault contact with each other. In fact, the faults that separate them include the major faults (or fault zones) of the Coast Ranges.

The rocks of the Franciscan core complex are typified by those that outcrop in the vicinity of San Francisco. Andrew Lawson, an influential geologist associated with the University of California at Berkeley, named these rocks the Franciscan Series, the name being taken from San Francisco itself. The area over which the Franciscan extends is much greater than the granitic-metamorphic core complex, and much of the northern part of the Coast Ranges province consists of Franciscan rocks.

Although much of the material in the Franciscan consists of sediment deposited in water, we should not think of the Franciscan as we ordinarily might think of a sequence of sedimentary rocks. Most sedimentary sequences are distinctly bedded and the units may be traced laterally for some distance. Not so with the Franciscan. It is a disorderly sequence in which distinct bedding is only locally present and in which it is commonly difficult or impossible to trace units laterally.

Among the most common rock types in the Franciscan are sandstones termed graywackes. The Franciscan graywackes are quite different from ordinary sandstones. We commonly think of a sandstone as consisting of rounded, distinct grains of sand that have been cemented together. In other words, ordinary sandstone looks like sand cemented to form rock. Franciscan graywacke, however, is a "dirty" gray or greenish gray. Furthermore, while many of the grains that make up the rock are the size of sand grains, they do not consist of rounded grains, but instead consist of a jumble of angular feldspar, quartz, and rock particles (of chert, shale, schist, and altered volcanic rock termed greenstone). The particles range widely in size (and may thus be said to be poorly sorted by size). Some of the finest particles consist of mica and chlorite. The greenish tinges are due to the presence of chlorite, which is a gray-green color. There is a complete gradation between sand and silt-sized particles, so that the distinction between sandstone and siltstone is not particularly meaningful. Instead, the term graywacke is a useful descriptive term, denoting the drab grayness of the rock. In addition to the sandstone, the Franciscan includes a great deal of dark mudstone.

The Franciscan sediments described above were deposited under conditions that are perhaps typical of the margins of continents that are undergoing active deformation. The graywacke sediments are "immature" in that they were probably derived by rapid erosion and were deposited swiftly without being moved about extensively by wave action on a former continental shelf. The transporting mechanisms evidently involved mass movement of material and flow of turbidity currents down submarine slopes.

Much of the material in the Franciscan is more or less directly related to volcanic sources. There are widespread occurrences of dark greenstones that represent dark fine-grained volcanic material that has been extensively altered. Much of the volcanic material may have been formed through submarine eruptions in which lava flows and blankets of tuff and volcanic breccia poured out. In places, the lava flows have distinctive "pillow" structure, which provides almost conclusive evidence of submarine eruptions. The pillows form when the hot lava comes in contact with seawater. Furthermore, some of the greenstone has formed as a result of the chemical interaction between the volcanic materials and the sodium ions in the seawater. Beds of chert and siliceous siltstone and mudstone are commonly associated with the submarine lava flows. For additional information on the lithology and chemical composition of the Franciscan, consult the report by Bailey, Irwin and Jones (1964).

The materials that make up the Franciscan can be classified according to two contrasting modes of origin. One type consists of the sediment (mudstone, siltstone, and graywacke) that was derived from land erosion and subsequently deposited near the margins of the continent. The other type consists of the submarine lava flows and the associated chert and siliceous mudstone that formed far out in the ocean. These two dissimilar groups of rocks were brought together and mingled mechanically by "ocean-floor spreading" in which movement of large segments, or "plates" of the earth's crust occurred, the oceanic plates moving against and sliding beneath the continental plates. The mingling of materials formed on the oceanic plate (lavas and chert) with those formed adjacent to the continent (graywacke, siltstone and mudstone) is probably responsible for the immense complexity of the Franciscan.

The basement complexes, particularly the Franciscan, are associated with ultrabasic rocks which consist mostly of serpentine. The origin of these masses of serpentine is not fully understood, but it is possible that they represent fragments of the earth's mantle (the mantle lies beneath the crust) which were incorporated as part of the Franciscan during movement of the oceanic plate beneath the continental plate.

Discussion of the details of ocean floor spreading and closely related concepts of continental drift are beyond our scope here, but they are well summarized in the Scientific American book *Continents Adrift* (1972).

## POINT REYES FIELD TRIP

The Point Reyes field trip provides a good opportunity to observe the contrast between the two basement complexes of the Coast Ranges. The route (Figure 1-1) of the field trip begins at the north side of the Golden Gate Bridge and extends northwestward to Point Reyes, following Highway 1 to the village of Point Reyes Station, and then turning off toward the Point Reyes peninsula.

The general geology of the region is represented by the simplified geologic map of Figure 1-2. The San Andreas fault zone forms the boundary between two basement complexes, namely that of Point Reyes peninsula itself, which lies southwest of the fault zone, and the remainder of the region on the opposite side of the fault zone. The geologic situation is similar in the Santa Cruz Range. Northeast of the zone, Franciscan rocks form the basement complex, whereas southwest of the fault zone, granitic rocks of the Sur Series form the basement.

### Stop 1: Fort Baker

The Fort Baker Military Reservation, which lies at the north end of the Golden Gate Bridge, provides an outstanding opportunity to observe bedded cherts and altered greenstones in the Franciscan assemblage. Actually there is a series of closely spaced localities at Stop 1, as denoted in Figure 1-3 (labeled 1A to 1C). These localities may be reached by leaving Highway 101 at the turnoff to Sausalito, and then descending to Fort Baker itself. Although Fort Baker is an active military establishment, the public is permitted to enter freely. Follow the road which winds past a number of buildings and then passes beneath the approaches of the Golden Gate Bridge. At this point, a curving segment of the road climbs steeply until it reaches a commanding position overlooking the Golden Gate itself.

All the localities display similar features, but because of their close proximity, it is worth walking from Stop 1A to Stop 1C. The bedrock exposed in the road cuts consists mostly of complexly folded beds (Figure 1-4) of reddish-brown chert and siliceous siltstone. The masses of layered chert occur in conjunction with masses of altered greenstone, which may be observed between Stops 1A and 1B, and elsewhere in the cuts along the road.

The origin of folding is somewhat conjectural, but it seems likely that at least some of the folding may have occurred shortly after the chert and siliceous siltstone layers were deposited. The Franciscan with its vast complexities is interpreted as having been deposited rapidly, and it may have undergone more or less continuous deformation as it accumulated.

A variety of other types of geological observations can be made at Stop 1. The intricate topography of coves and headlands formed by the steep slopes rising above the sea is not only scenically attractive, but also reflects the influence of landslides in shaping topography. The undercutting effects of wave action are partly responsible for the steep slopes. Wave attack tends to be concentrated on the headlands that separate the coves. The patterns of the waves, particularly their refraction in conforming to headlands and coves, provide interesting observations, depending on the intensity of the surf and the angle of the sun's rays.

## Stops 2 and 3: Sea Cliffs

From Stop 1, return to the main highway (combined U.S. 101 and California 1) and proceed northward until Highway 1 branches from U.S. 101. Proceed west, and then northwest on Highway 1, toward Stinson Beach until you reach Stop 2 (the locations of stops are shown in Figure 1-5).

Stops 2 and 3 provide fine opportunities to observe the effects of erosion on a coastline that has undergone repeated uplift. In places, remnants of wave-cut terraces lie at relatively high elevations above sea level (Figure 1-6). The rock exposed in the sea cliffs and roadcuts consists of a mixture of lithologies in the Franciscan. Gray siltstone is prominent, but in places, relatively small masses of intrusive serpentine crop out, as at Stop 2. The serpentine is a glistening light greenish gray, with a slick, "serpent-like" feel that is characteristic of some forms of serpentine. In places, bedded cherts, or siliceous siltstones, stand out as resistant masses, as at Stop 3. Some of the chert contains minute spots, which probably represent the altered remains of radiolaria. Radiolaria are microscopic floating organisms with skeletons composed of silica that flourish in seawater rich in dissolved silica. A close relationship between submarine volcanism and the presence of cherts or siliceous siltstones containing radiolaria has been noted in other regions. The reaction of hot igneous rocks with seawater was a probable source of the dissolved silica.

## Stops 4 and 5: Bolinas Lagoon

Stop 4 overlooks Bolinas Lagoon and the curving bay mouth bar that separates the lagoon (Figure 1-5) from the open sea. The bay mouth bar is a textbook example and illustrates the role of longshore currents in building a sandbar that largely encloses the lagoon. The surface trace of the San Andreas fault crosses the western end of the sand bar. Movement occurred along the fault during the San Francisco earthquake of April, 1906. Bolinas Lagoon itself is closely related to the topographic influence of the San Andreas fault zone, the local trace of the fault being marked by a straight, elongate valley which has been "drowned" at each end, creating Bolinas Lagoon on the southeast and Tomales Bay on the northwest (Figures 1-1 and 1-2).

Stop 5 is along Highway 1 where it lies adjacent to Bolinas Lagoon. At low tide, the intricate complex of mudflats and tidal channels can be observed. The twice daily tidal movement of seawater in and out of Bolinas Lagoon maintains the continuity of tidal channels, and of course, has strong influence on the environmental properties of Bolinas Lagoon.

8 • Coast Ranges Province

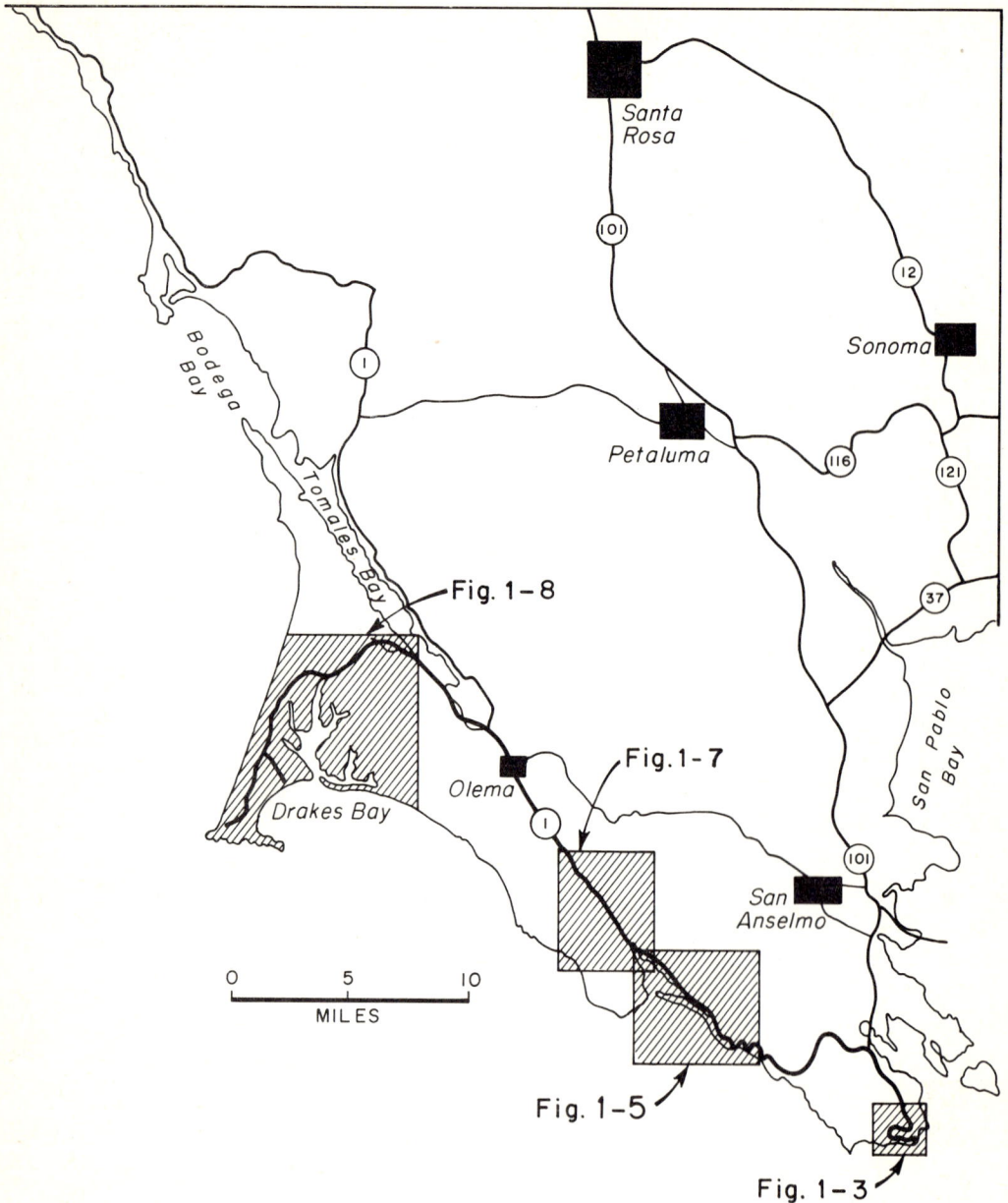

**Figure 1-1.** Map showing route and stops of Point Reyes field trip. Diagonally ruled areas denote locations for which route maps (identified by figure numbers) provide details of field trip stop localities.

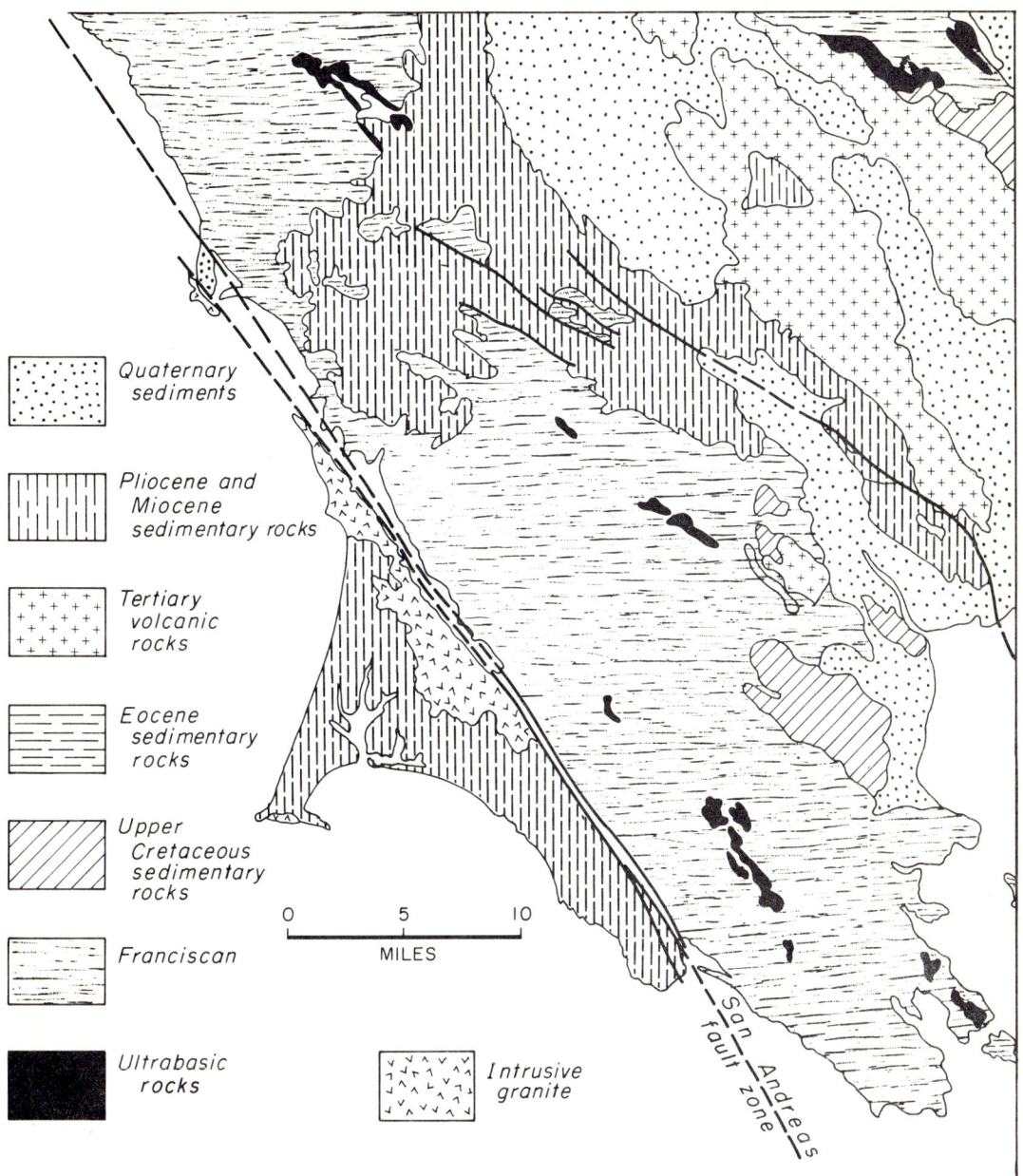

**Figure 1-2.** Simplified geologic map of area of Point Reyes field trip. Area of map coincides with Figure 1-1.

# 10 • Coast Ranges Province

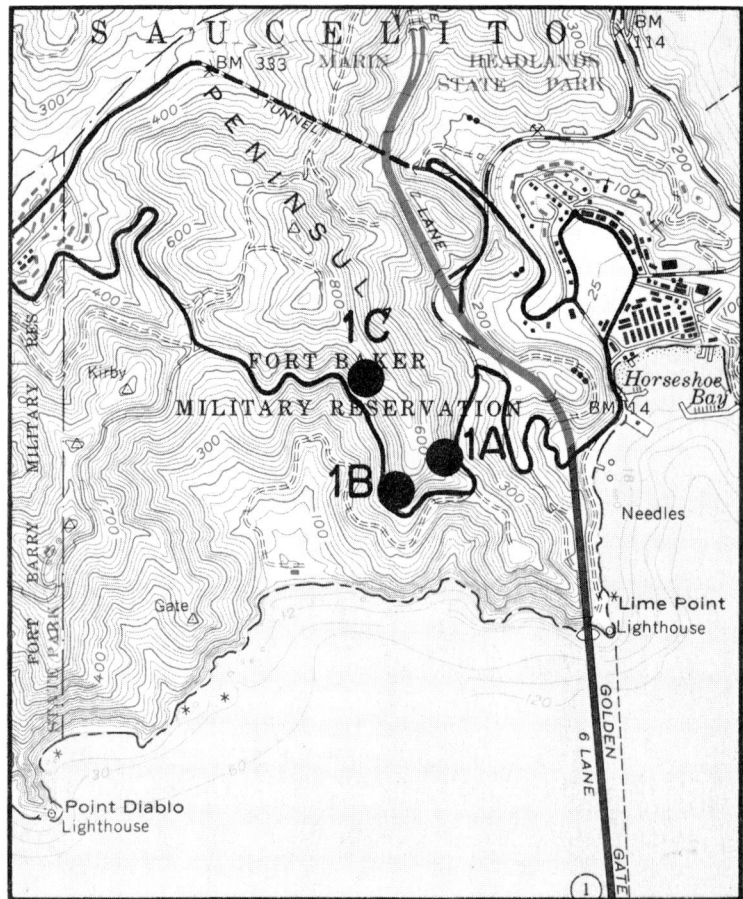

**Figure 1-3.** Field trip route and stops on north side of Golden Gate. East-west width of map is 1.5 miles. (From U.S.G.S. San Francisco North 7½' topographic quadrangle.)

**Stop 6: Fault Zone Topography**

Highway 1 follows the valley whose linear elongation is related to continued movement along the San Andreas fault zone. This movement has created a number of topographic paradoxes, as for example, the presence of two streams that are essentially parallel to each other and lie only about one third of a mile apart, and yet flow in opposite directions. Inspection of Figure 1-7 reveals that Pine Creek flows southeastward into Bolinas Lagoon, whereas Olema Creek flows northwestward into Tomales Bay. The trace of movement along the fault that accompanied the 1906 earthquake lies close to the two streams. Highway 1 follows the fault zone for some distance.

There are a large number of features in the vicinity related to movement along the San Andreas fault during the 1906 earthquake. The maximum displacement of the fault during the earthquake occurred at Olema, the crust of the earth on the southwest side of the fault moving about 21 feet toward the northwest relative to the crust on the other side of the fault. At Point Reyes National Seashore headquarters in the village of Point Reyes Station, there is a "San Andreas fault walk" which contains a number of exhibits, including the trace of movement along the fault in 1906. For additional information, consult the book by Iacopi, which contains a detailed description of the effects of displacement along the San Andreas fault zone accompanying the 1906 earthquake, and the report edited by Oakeshott (1959). From Point Reyes Station, turn west and thence northwest through Inverness, and then take the road to Point Reyes.

**Stop 7: Granitic Basement Complex**

Stop about 500 feet west of the point where the highway leading to Point Reyes leaves the shore of Tomales Bay (Figure 1-8). Look for granitic rock in the roadcut on the north side of the highway. The granitic rock is representative of the basement complex southwest of the San Andreas fault zone, contrasting with the Franciscan which we have observed at Stops 1 to 4. The granitic rock is also exposed in the sea

**Figure 1-4.** Intricately folded layers of chert and siliceous siltstone exposed at Stop 1. Height of cliff is about 25 feet.

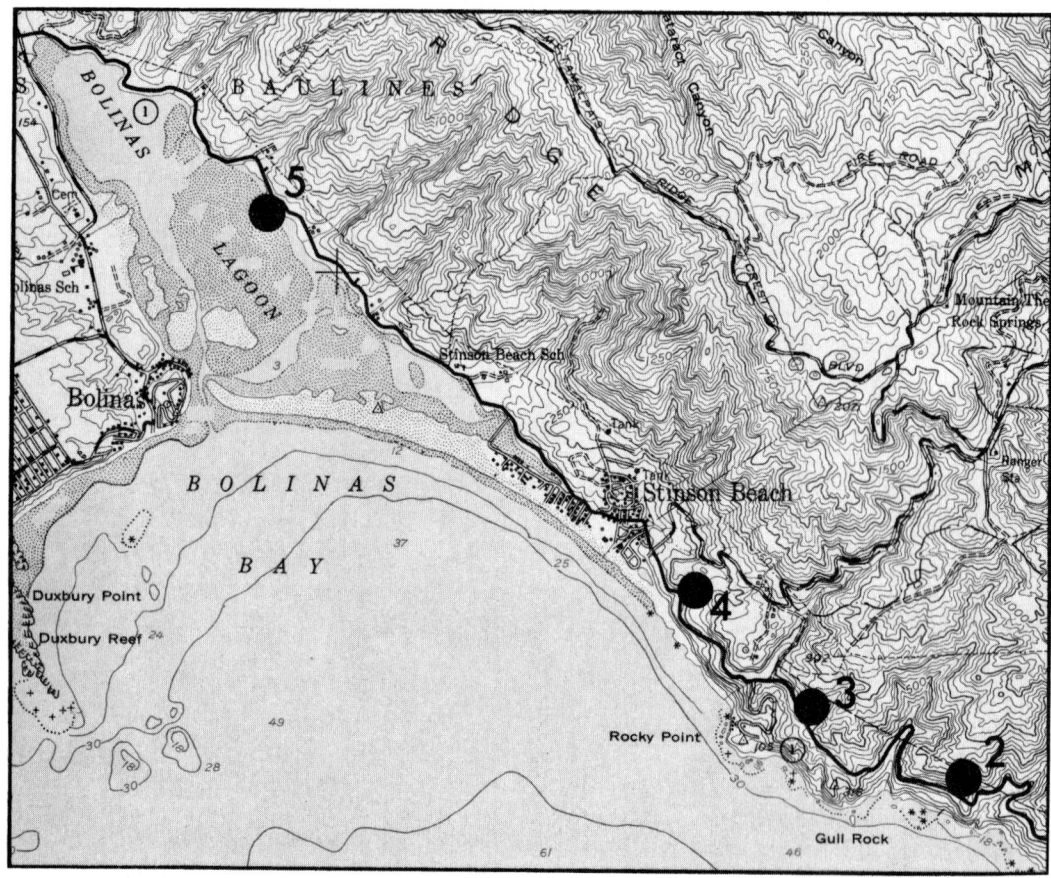

**Figure 1-5.** Location of Stops 2 to 5 along Highway 1. East-west length of map is 5.5 miles. (From U.S.G.S. Mt. Tamalpais 15' topographic quadrangle.

cliff at Point Reyes itself. Tertiary strata, which overlie the granite rock, form a broad syncline in the area between Stop 7 and Point Reyes.

**Stop 8: Drakes Estero**

Drakes Estero (Figure 1-8) is a classic example of a drowned tree-like drainage system. The various arms of Drakes Estero are the tributary valleys. Drakes Estero (in common with other bays, estuaries, and lagoons of the region, including San Francisco Bay and Bolinas Lagoon) have been flooded by the rise of sea level accompanying the continued melting of glaciers occurring at the close of the Pleistocene Epoch. Evidence from other parts of the world indicates that sea level is still rising. Thus, the flooding of bays and estuaries in the region is still in progress. Drakes Estero, like Bolinas Lagoon, is shallow and consists of a complex of tidal channels and tidal flats. The tidal flats are exposed at low tide.

Aquatic plants contribute substantially to the estuary filling process. It is instructive to walk out on the dry part of the marsh land (between the road and the tidal flat) and jump up and down. The marsh is very springy underfoot, which reflects the properties of the mixture of plant material, silt, and mud, which forms the deposits that are filling the marsh.

### Stops 9 and 10: Point Reyes Beach

Point Reyes Beach is an exciting place. The surf is exceptionally strong and there usually is a strong wind from the sea toward the land. The broad sandy beach (Stop 9) is, in part, an effect of the longshore currents in furnishing a supply of sand along the coastline. The sand on the beach immediately adjacent to the strandline is heaped into a regular series of cusps. If you walk along the beach you will be conscious of the small-scale roller coaster effect, the difference in elevation from bottom to top of an individual cusp being several feet.

The winds are responsible for carrying sand from the beaches and heaping it into dunes on the land in back of the beach (Stop 10). The peculiar topography (Figure 1-8) adjacent to the beach reflects the elongate shapes of the sand dunes, which

**Figure 1-6.** View along coastline in vicinity of Stops 2 and 3, looking northwest. Note numerous marine stacks which project above sea level and rest on wave-cut bench that is presently forming. Nearly horizontal segment of skyline is remnant of uplifted wave-cut bench.

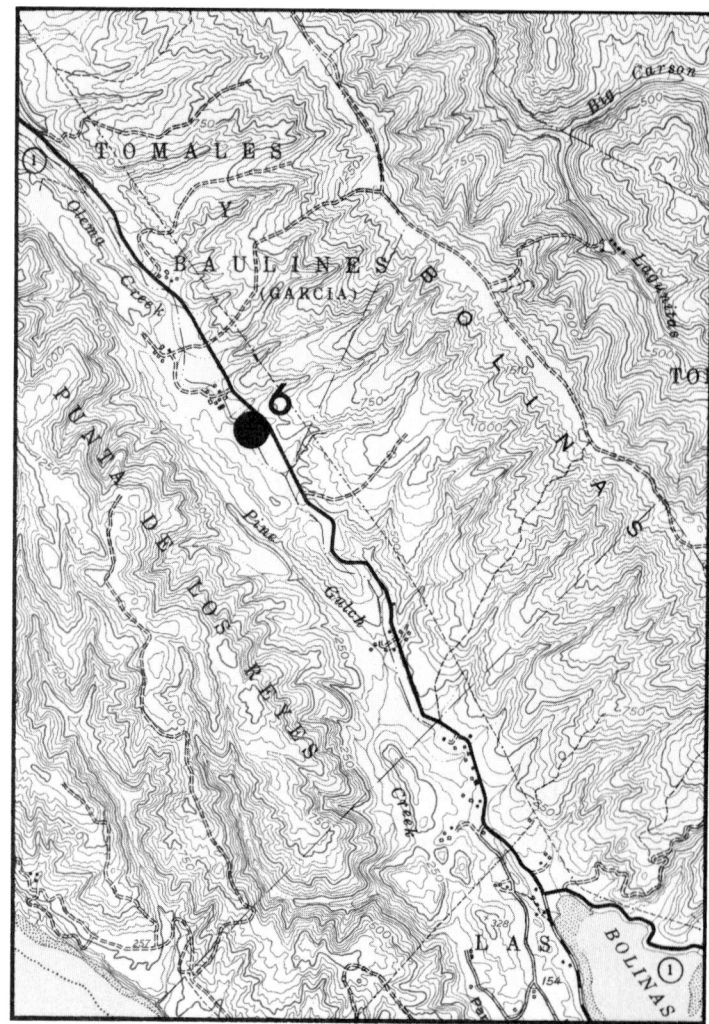

**Figure 1-7.** Location of Stop 6 in San Andreas fault zone northwest of Bolinas Lagoon where parallel streams flow in opposite directions. East-west width of map is 3.7 miles. (From U.S.G.S. Mt. Tamalpais 15' topographic quadrangle.)

result from the transport of sand from the beach by wind.

### Stop 11: Drakes Beach

Drakes Beach is a center of activity for the Point Reyes National Seashore, and is protected from heavy surf by Point Reyes, which forms a partial enclosure creating Drakes Bay (Figure 1-8). The cliff in back of Drakes Beach consists of light gray-buff siltstone of Pliocene age which occurs in distinctive beds. Although the beds are nearly horizontal (Figure 1-9), they form part of a broad syncline extending between Point Reyes and the granitic rock outcrop at Stop 7. In places, vertical faults with a displacement of a few feet have offset the beds. The beds are fossiliferous and contain fossils as diverse as starfish and whale bones. The siltstone has been highly fractured and alteration of the rock adjacent to fractures causes the fracture traces to stand

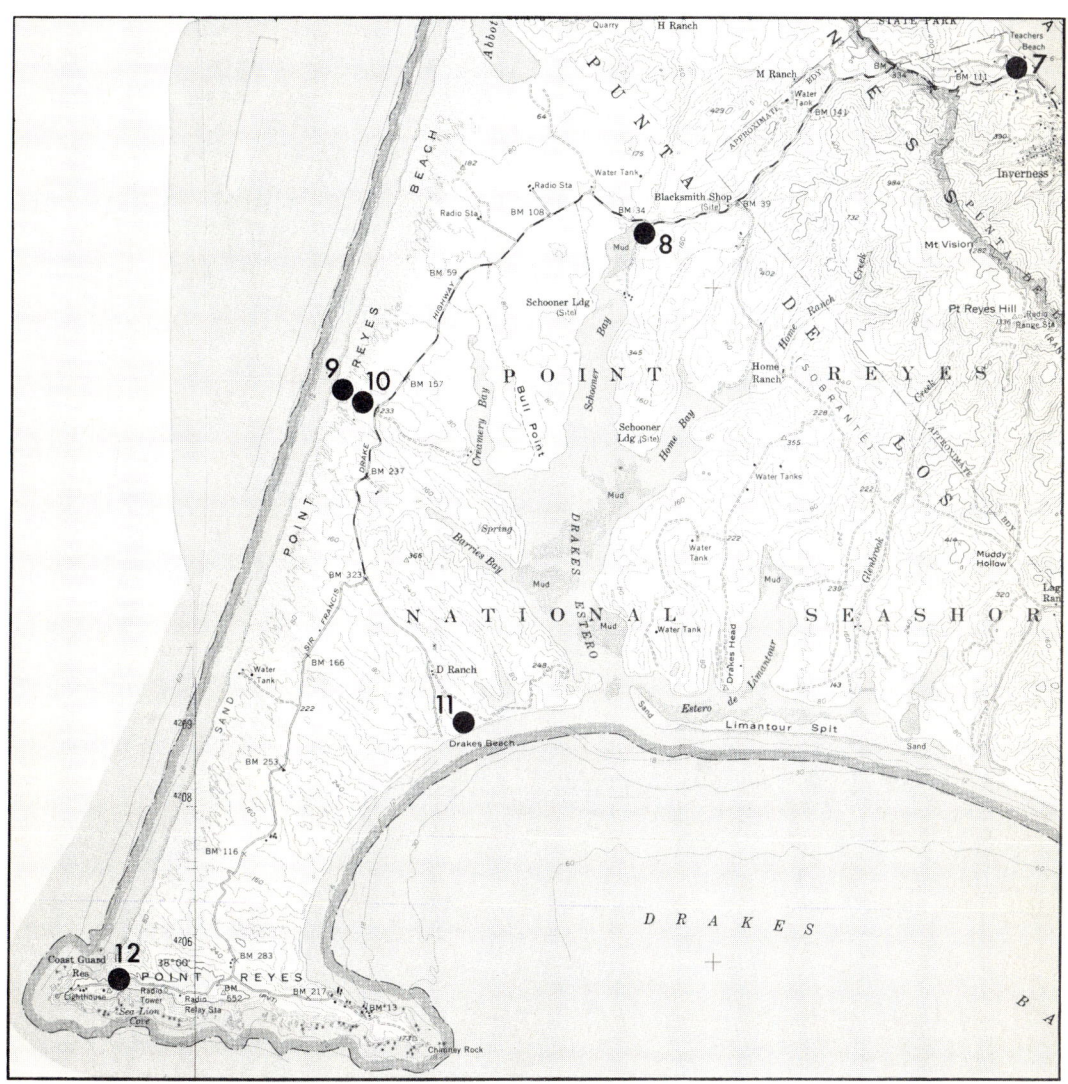

**Figure 1-8.** Southern part of Point Reyes peninsula showing location of field trip stops 7 to 12. East-west width of map is 9.2 miles. (From U.S.G.S. Point Reyes 15′ topographic quadrangle.)

**16** • COAST RANGES PROVINCE

**Figure 1-9.** Drakes Beach, looking east.

out in relief, particularly along the upper surfaces of bed exposed at low tide.

### Stop 12: Point Reyes

Point Reyes itself offers the most spectacular scenery of the entire Point Reyes peninsula. Point Reyes is formed by a mass of granitic rock which is overlain by conglomerates of early Tertiary age. The conglomerates contain a varied assemblage of cobbles and pebbles, including some that were undoubtedly derived through local erosion of the granitic rock.

### BIBLIOGRAPHY

BAILEY, E. H., ed. 1966. *Geology of Northern California.* California Div. Mines and Geology. Bull. 190.

Symposium volume which contains thirty-four papers dealing with the geology of northern California. Most of the papers are at the technical level and directed to professional geologists. Six field trip guides are included. 507 pp.

BAILEY, E. H.; IRWIN, W. P.; and JONES, D. L. 1964. *Franciscan and related rocks, and their significance in the geology of western California.* California Div. of Mines and Geology. Bull. 183.

Detailed description of composition of Franciscan Formation over its broad extent in the Coast Ranges province. 177 pp.

CURTIS, G. H.; EVERNDEN, J. F.; and LIPSON, J. I. 1958. *Age determination of some granitic rocks in California by the potassium-argon method.* California Div. of Mines. Spec. Rept. 54.

Includes radiometric age determinations of the granitic rocks in the Coast Ranges province. 16 pp.

DICKINSON, W. R.; and GRANTZ, A. 1968. *Proceedings of Conference on Geologic Problems of San Andreas Fault System.* Stanford Univ. Pubs. in Geol. Science. vol. 11.

Symposium volume containing forty-seven papers that deal broadly with the San Andreas fault system and other aspects of California geology that involve consideration of the San Andreas fault's history. 392 pp.

HIGGINS, C. G. 1961. San Andreas fault north of San Francisco, California: *Bull. Geol. Soc. Amer.* 72: 51-68.

Detailed discussion of nature of fault zone, including its width, and amount of vertical and horizontal displacement since the middle Pliocene.

IACOPI, R. 1964. *Earthquake country.* Menlo Park, Calif.: Lane Magazine & Book Co.

Vividly illustrated book that authoritatively treats the routes of the San Andreas fault and other major faults in California. Contains numerous photographs of topographic features related to the faults, and scenes of destruction wrought by past earthquakes caused by abrupt movement along these faults. 192 pp.

OAKESHOTT, G. B., ed. 1959. *San Francisco earthquakes of March 1957.* California Div. Mines Spec. Rept. 57.

Describes results of 1957 earthquakes, and also includes an extensive discussion of the San Andreas fault zone. 127 pp.

OAKESHOTT, G. B. 1971. California's changing landscapes: New York: McGraw-Hill Book Co.

A superbly written volume by a geologist with an intimate knowledge of California's geology.

The book also includes a general introduction to the principles of geology which are interwoven with the descriptions of California's geology. 388 pp.

Scientific American 1972. *Continents adrift.* San Francisco: W. H. Freeman & Co.

A collection of articles published in the magazine Scientific American between 1952 and 1972 that deal with continental drift, seafloor spreading, and plate tectonics. 172 pp.

SHARP, R. P. 1972. *Geology field guide to southern California.* Dubuque, Iowa: Wm. C. Brown Company Publishers.

Companion volume for southern California in the Wm. C. Brown *Regional Geology Series.* 181 pp.

SHELTON, J. S. 1966. *Geology illustrated.* San Francisco: W. H. Freeman and Co.

Superbly illustrated volume which stresses the visual observation approach to physical geology. Strongly recommended to users of the *Regional Guide Series.* 434 pp.

TALIAFERRO, N. L. 1943. *Geologic history and structure of the central Coast Ranges of California.* California Div. Mines Bull.

Treats general structure of Coast Ranges. 118 pp.

WEAVER, C. E. 1949. Geology of the Coast Ranges immediately north of the San Francisco Bay region, California. *Geol. Soc. America Mem.* 35.

Regional geology of area north of San Francisco Bay area. 242 pp.

# Chapter 2

# Coast Ranges: Santa Cruz Mountains Coastal Area Field Trip

## GENERAL GEOLOGY OF THE SANTA CRUZ MOUNTAINS

The field trip routes are shown in Figure 2-1 and the general geology of the Santa Cruz region is portrayed on the simplified geologic map of Figure 2-2. The patterns defining the two basement core complexes are identified in the lower left of Figure 2-2. Granitic igneous rocks are distinguished from metamorphosed sedimentary rocks in the granitic-metamorphic core complex. The other basement complex consists of the Franciscan and the ultrabasic rocks that are associated with it. Ultrabasic rocks are rich in magnesium and iron and contain less silica than granitic rocks. In general, the ultrabasic rocks have been altered to serpentine. The map units that are younger than the basement complexes are identified in the upper right part of Figure 2-2. These map units consist of layered sedimentary rocks, volcanic rocks, and unconsolidated sediments. These units rest upon and obscure the basment complexes in much of the area.

In examining the map, your attention should be drawn to the San Andreas fault and to several adjacent faults whose surface traces over the land are more or less parallel. Collectively these closely related faults form a zone that divides the Santa Cruz Mountains into two geologic subprovinces that have important differences. On the southwest side of the zone, the granitic-metamorphic core complex is present, whereas on the northeast side, the Franciscan core complex is present and the granitic-metamorphic complex is excluded. The San Andreas fault zone, viewed broadly, establishes the boundary between these geologic subprovinces.

## MONTARA MOUNTAIN AREA

The area from Point San Pedro to Moss Beach provides some of the most spectacular scenery of the Santa Cruz Mountains. It also contains some exceedingly interesting geologic features that are well exposed. If you are a newcomer to the geology of the Santa Cruz Mountains, it is an excellent locality to begin your geological exploration. The locations of the eight field trip stops in the vicinity are shown in Figure 2-3.

The dominant topographic feature in this area is Montara Mountain, which rises directly from the sea, extending to a maximum elevation of 1,898 feet above sea level. Geologically, Montara Mountain consists mostly of a large mass of granodiorite which is part of the granitic-metamorphic core complex. Our interest is drawn not only to the granodiorite, but also to the sedimentary rocks that are in contact with it. The sedimentary rocks have been complexly deformed by folding and faulting. Furthermore, erosion of the granodiorite in the past has provided a source of material incorporated in some of the sedimentary rocks.

**Stop 1: Point San Pedro**

Point San Pedro is reached by driving to Shelter Cove and walking the remaining distance along the path at the foot of the cliff. Shelter Cove is a private beach and the owners sometimes exact a small charge for parking. The rocks from Shelter Cove to Point San Pedro consist of alternating beds of graywacke sandstone, siltstone, and mudstone. In places, the strata consist of thin layers from which more resistant sandstone layers stand out in relief on weathered surfaces. The beds generally dip steeply. The strata are particularly well exposed on San Pedro Rock, which is an island that can be reached by foot at low tide. (Figure 2-4).

Our principal quest at this stop is to observe these strata, reflecting on their composition, and the conditions under which they were deposited, their subsequent deformation and uplift, and their progressive destruction by erosion that is now taking place. The strata are composed of an almost rhythmic alteration of coarser and finer material (sand and silt-sized particles). If we look closely, some of the silty layers are irregularly mottled due to the activity of burrowing organisms when the sediments were still soft. When the individual layers are viewed from a distance, they appear uniform in thickness, but when examined at close range (Figure 2-5), variations in thickness can be seen. Furthermore, close examination reveals the presence of small scale cross laminations in which groups or sets of individual laminae are inclined with respect to each other. These sedimentary details are only of modest interest, in themselves, but their importance lies in the fact that they provide some of the evidence from which we can attempt to interpret the depositional history of the strata. We might interpret the strata as deposited on a submerged topographic shelf lying some miles from shore, with water depths measured in hundreds or several thousand feet. Periodically, clouds of fine sand and silt suspended in water probably were spread over the shelf, forming the layers of sediment. Vigorous wave action did not affect the sediments as they were deposited, judging from the thin beds and their delicate laminae. On the other hand, there is evidence that gentle currents were effective in depositing the materials, producing the cross laminations which are generally regarded as an indication of deposition by currents.

Other sedimentary features that are of interest include the conglomerates containing a jumble of pebbles and cobblestones that have been cemented together in a matrix of siltstone. Some of the cobble-sized fragments are very angular. The best opportunity to observe the conglomerates is with-

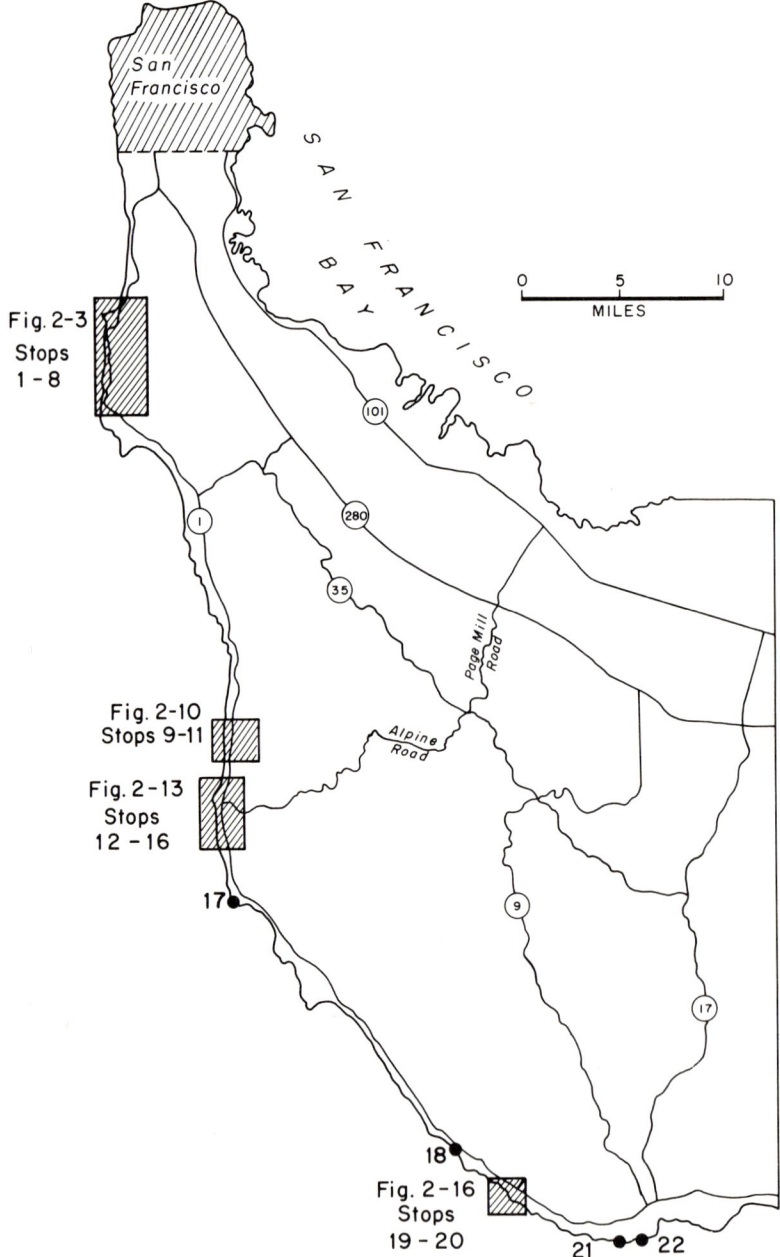

**Figure 2-1.** Outline map of San Francisco Peninsula and northern Santa Cruz Mountains region showing access routes to field trip stop localities. Heavy outlines indicate boundaries of large-scale maps (which are identified by numbers) of other illustrations which give details of field trip stop locations. Field trip stops not shown on detailed maps are identified by solid black circles.

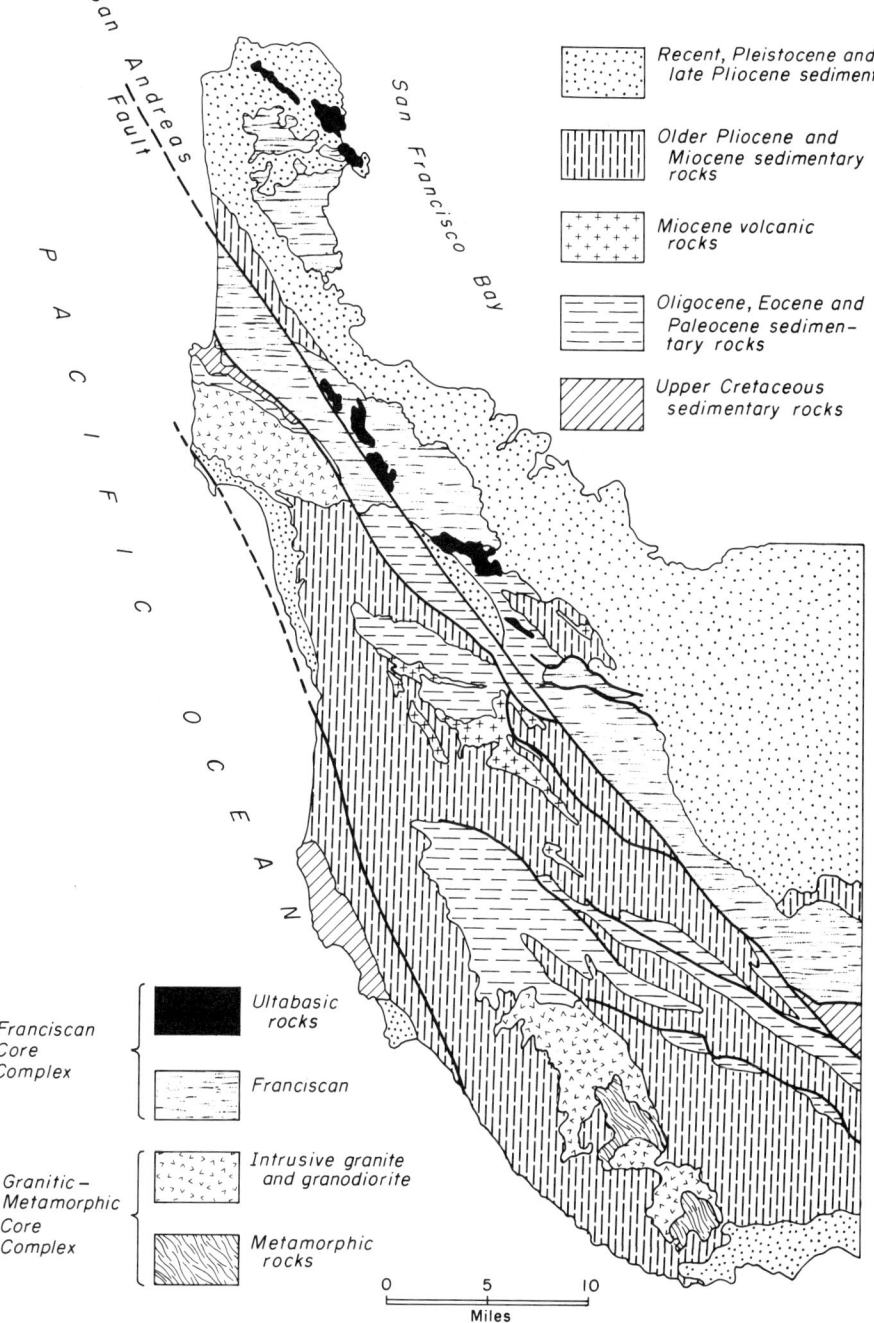

**Figure 2-2.** Simplified geologic map of San Francisco peninsula and northern Santa Cruz Mountains.

22 • Coast Ranges

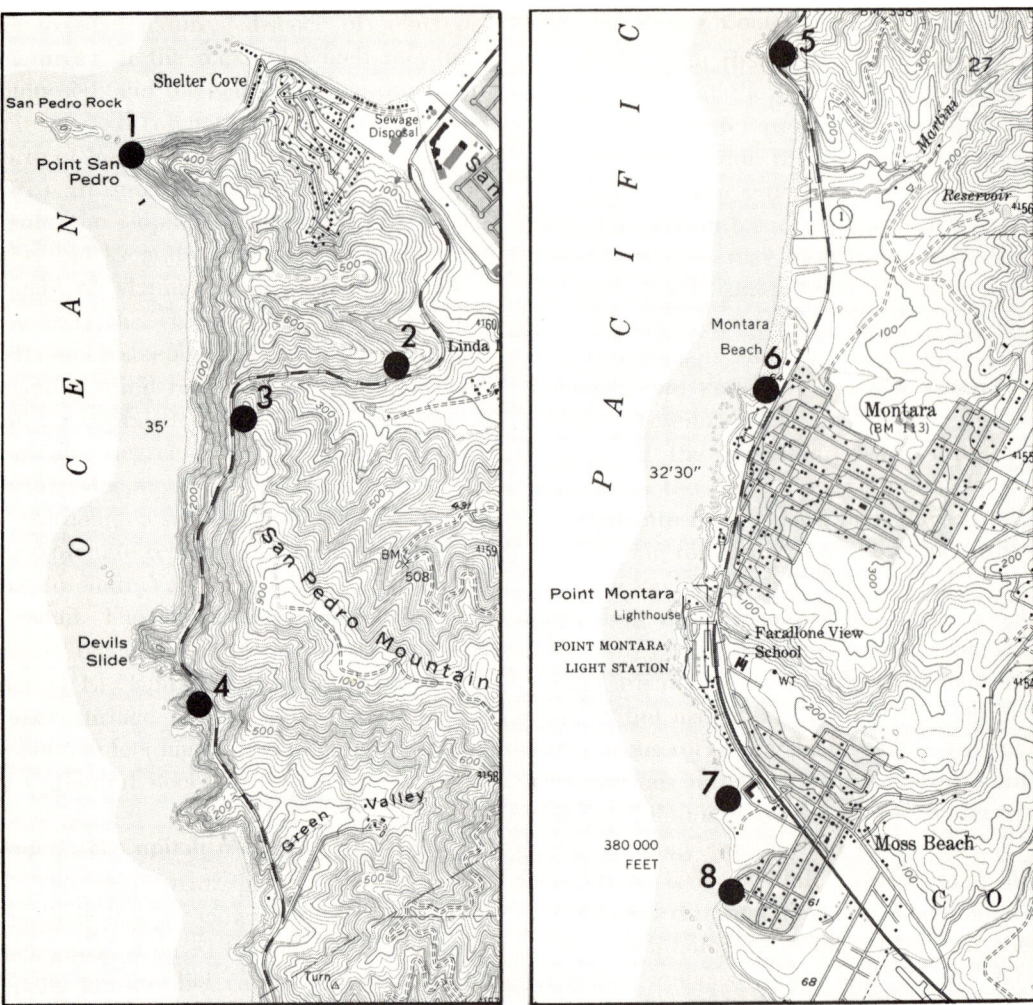

**Figure 2-3.** Maps showing field trip stops in Montara Mountain area. Map segment on left is immediately north of segment on right. Width of each segment is 1.4 miles. (From U.S.G.S. Montara Mountain 7½′ topographic quadrangle.)

in the large loose boulders that lie immediately above the beach, because the boulders probably came from the conglomerates exposed in the cliff above.

When we visit Point San Pedro, we are acutely conscious of the sea. The steep cliffs, the boulder-strewn beach, and the jagged form of San Pedro Rock all represent the influence of wave activity. Here, as well as many other places along the California coastline, wave activity is carving a platform or bench that is shallowly submerged. As time passes, the cliffs tend to gradually retreat under the onslaught of wave activity, and the wave-cut bench is broadened. The material removed is carried out to sea and deposited as sediment. Shore processes are part of the ceaseless recycling of materials through erosion and deposition. While shore processes are important only along coasts (and thus affect only a minute proportion of the area of a continent), they tend to strongly influence the topography of coastal regions. San Pedro Rock, for example, is undergoing vigorous erosion, and has only a short remaining life.

## Stop 2: Conglomerate Exposure

Stop 2 is at a road cut on the north side of Highway 1 (Figure 2-3), where there are excellent exposures of a coarse conglomerate composed of large, rounded boulders of granodiorite. The boulders and cobbles are mostly composed of granodiorite which is now deeply weathered, but other materials, such as siltstone and quartz are present. The conglomerate is virtually unsorted, with sizes ranging from small pebbles to boulders three feet or more in maximum dimensions. Important questions here include the source of material for the boulders and cobbles, and the conditions under which the conglomerate was deposited. Possibly the conglomerate consists of material moved by a submarine slide which had accumulated earlier on a shallowly submerged marine shelf.

**Figure 2-4.** San Pedro Rock viewed from shore (looking west-northwest).

**Figure 2-5.** Bedding details of rocks exposed at Point San Pedro. Lighter layers consist of siltstone whereas darker layers are mudstone. Pen provides scale.

**Stop 3: Devil's Slide Area**

Continue on Highway 1 for 0.4 mile until Stop 3. Park at any convenient spot on the west side of the highway. Parking is discouraged in the Devil's Slide area, so you may wish to pause only briefly at Stop 3.

The Devil's Slide area provides a spectacular view seaward. Our interest is drawn to the thin-bedded graywacke siltstone and mudstone exposed in the cliffs. In places, these strata are folded intricately and some of the beds are overturned. While the folds appear similar to those resulting from structural deformation of the earth's crust, there is strong suspicion that much of the folding has resulted from landslide activity on the steep slopes. The highway at Devil's Slide is famous for landslides and has undergone repeated repairs after slides have occurred.

**Stop 4: Sea Cliffs**

Proceed southward from Stop 3 for about 0.8 mile. Watch closely for a wide shoulder area on the west side of the highway immediately after you travel through a deep road cut. About 600 feet before you reach Stop 4, there is a short driveway, followed by a steep stairway and footpath on the west side of the road. Stop 4 is just beyond the second road cut after you have passed the stairway. When you reach Stop 4, look for a footpath that leads westward for about 200 feet. The path is poor and is flanked by abundant poison oak. Stop on the path when you have gone far enough to get a good view of the sea cliffs.

The cliffs are carved from the Montara Granodiorite, which is a coarse-grained granitic rock with feldspar crystals one-fourth of an inch or more long. A radiometric date of the Montara Granodiorite places its age at roughly 90 million years, although it may be older. The granodiorite is intersected by a network of joints as well as narrow intrusive light-colored dikes. If you look at the steep cliff on the other side of the cove, toward the northwest, you can see a number of dikes. Note also the intensity of the surf at the foot of the cliffs. The system of joints in the rock, plus intense wave attack, have combined to produce exceedingly rugged topography.

**Stop 5: Montara Granodiorite**

Proceed southward for about one mile from Stop 4. Park on the wide shoulder on the west side of the highway. From here you obtain an excellent view (Figure 2-6) of Montara Mountain toward the north. Note the several remnants of wave-cut terraces that are now much above sea level and which form topographic spurs in the middle foreground, as well as in the distance.

Stop 5 also provides a good opportunity to examine some of the features of the Mon-

**Figure 2-6.** Montara Mountain viewed from south (Stop 5). Note remnants of uplifted wave-cut terraces in middle foreground.

tara Granodiorite. Granodiorite is an igneous rock that is closely related to granite. In fact, it would be only a minor error to describe it as granite because it exhibits gradations from granite to granodiorite from place to place, granodiorite having a little less quartz than granite. One of the interesting features of the rock is an abundance of large feldspar crystals which are up to an inch or more in longest dimensions. The rock consists of a mixture of plagioclase feldspar (i.e., sodium-calcium feldspar), orthoclase (potassium feldspar), quartz and biotite. The rock is deeply weathered in most places, crumbling to coarse sand.

The Montara Granodiorite is named after Montara Mountain. We do not know the full extent of the Montara Granodiorite because it is covered by younger rocks to the north and south. Along the northeast side, the granodiorite is separated from the Franciscan by a fault (Figure 2-2). The Montara Granodiorite is one of a number of large intrusive masses that form part of the granitic-metamorphic core complex. For example, the granitic masses that crop out 30 miles toward the southeast are broadly related to the Montara Granodiorite, but they represent different intrusions.

The Montara Granodiorite is not a uniform mass, formed in a single act of intrusion. Internally the granodiorite mass shows a complex history. Dikes consisting largely of coarsely crystalline feldspar and quartz (Figure 2-7), have invaded the granite, either displacing it or replacing it. The dikes are necessarily younger than the main mass of granodiorite, although it is possible and perhaps probable that they are only very slightly younger, representing a late stage in the overall intrusive process. Some of the dikes are more or less sheetlike, whereas others pinch and swell, changing direction abruptly.

In addition to providing a good exposure of the granodiorite, the roadcut also exposes an overlying marine terrace deposit that rests upon a wave-cut bench. The terrace sedimentary material erodes differently than the weathered granodiorite and the boundary between the terrace deposits and the granodiorite is obvious. The terrace at this locality is lower (and therefore younger) than the terraces observed in the distance at Stop 5. The succession of terraces provides strong evidence of repeated uplift of the area. Changes in sea level have also played a role in the origin of these features.

**Stop 6: Montara Beach**

Proceed about 0.8 mile south of Stop 5 until you reach Montara Beach. Montara Beach is a public beach and is part of the San Mateo County State Beach Park. Leave your car in the parking lot (marked by a sign) on the west side of the highway and walk down the short, steep road to the beach itself.

Our interest here is focused on the terrace deposits that lie above the beach. The highway crosses the terrace after descending from Montara Mountain. The cliffs above the beach provide good exposures of the terrace deposits, and also of the underlying bedrock on which the deposits were laid. The deposits consist of layers of sand and fine gravel that are partly consolidated, and exhibit large-scale cross stratification in which the layers are inclined at angles of 20° to 30° from the horizontal. The sandstone deposits rest unconformably on the irregular surface of the Montara Granodiorite.

**Figure 2-7.** Roadcut on east side of highway at Stop 5 revealing dikes that have intruded granodiorite. Dikes in lower central part of picture are offset by fault that has displaced rock below fault toward right. Marine terrace deposits at top of cut exhibit badlands-type erosion.

The loose sand at Montara Beach is worth examining closely with a hand lens. It consists mostly of coarse quartz and feldspar grains which are angular to slightly rounded. Comparison of the sand with the granodiorite suggests that the sand has probably been derived from the granodiorite, although some may have been derived from the terrace deposits which were supplied earlier with material by the decaying granodiorite.

**Stop 7: Moss Beach (A)**

There are two closely related observational stops at Moss Beach. To reach Stop 7, turn off Highway 1 and park opposite the parking area adjacent to a small motel (Dan's Motel) west of the highway. A narrow beach which lies at the foot of the slope is accessible at low tide. Several impromptu footpaths provide ready access to the beach which is part of the James V. Fitzgerald Marine Reserve. Do not go down to the beach at high tide or when the surf is high because of the danger. At other times, the beach is relatively safe, but be on guard against large waves that may inundate the beach with little warning.

This stop illustrates features somewhat similar to those of Montara Beach, but there are some surprises. If we examine the rock exposures (Figure 2-8), we find that there are two distinct sedimentary units. One is the upper terrace deposit consisting of

**Figure 2-8.** Exposure at Stop 7. Unit immediately above beach sand is cross-stratified conglomerate and coarse sandstone. Large boulders are part of overlying unit that consists of poorly consolidated terrace deposits lying above wave-cut bench. Total height of cliff is about 20 feet.

poorly consolidated sand and silt. The lower part of this unit, however, contains large, rounded boulders of granodiorite. The other unit consists of an intricately cross-bedded, ill-sorted but firmly consolidated jumble of conglomerate and sandstone. The beds dip steeply in places. The boundary between the units is very distinct, and consists of a wave-cut bench on which the overlying sediments have been deposited.

A geological surprise at Stop 7 is provided by observing the lower unit carefully, walking eastward for about 200 feet along the beach. At first, the lower unit is obviously sedimentary, consisting of cross-stratified coarse conglomerate and coarse sandstone undoubtedly derived through erosion of granodiorite. Suddenly, the lower unit changes to granodiorite which is in place. The challenge is to find the place where the transition occurs, which is obscured by a high degree of fracturing of the rock. A fault affects both the older rock and the overlying sediments and is thus very young.

The geologic history that can be deduced at Stop 7 is worth careful thought. It is clear that the two units are intimately involved with changes of the elevation of land with respect to sea level. Did the land rise or did the sea fall? The presence of much higher (and therefore older) terraces in the vicinity (as on Montara Mountain) strongly suggests that, in general, the land has risen relative to sea level. But, of the two units exposed in the cliff at Stop 7, the overlying unit (poorly consolidated sand and silt with big boulders near its base) is unquestionably younger than the unit below because it rests on the lower unit. Recall that the law of superposition decrees that the oldest sedimentary strata in a given sequence are progressively overlain by younger strata (except, of course, where the strata have been turned upside down by structural deformation).

The geologic processes at work at Stop 7 should be considered. The granodiorite is being vigorously eroded by the surf, yielding coarse sand that contains an abundance of feldspar and quartz grains. Loose cobbles and boulders of granodiorite are also being deposited at present, particularly in the surf. We can imagine the mixture of sand and boulders that are being deposited now at some slight distance from shore. The large boulders that are part of the terrace deposits (those imbedded near the base of the poorly consolidated sand and silt), were probably deposited under similar conditions—perhaps in heavy surf, not far from shore, with nearby exposures of granodiorite as a source of supply.

**Stop 8: Moss Beach (B)**

Stop 8 is only a short distance from Stop 7, but it is not feasible to walk directly along the shoreline between the two stops. Instead, drive through the residential part of Moss Beach, park at the parking lot that is part of the beach facility, and walk to the beach. There are a number of geological sights here. Our attention is drawn to the dark sandstones and conglomerates that are exposed at the beach. Walk north for several hundred yards from the ramp that descends from the parking area to the beach. If the tide is low, you will see gently folded strata that form broad folds (Figure 2-9) and stand out in relief because of the difference in erosional resistance of certain beds. The strata exposed near the beach and in shallow water are essentially equivalent to the sandstones and conglomerates in the lower unit at Stop 7. As at Stop 7 some of the sandstone beds are cross stratified. Most of the conglomerates are composed of pebbles and cobbles of granodiorite and sedimentary rocks. In places, granodiorite boulders are present. There can be little doubt that these strata were deposited under marine conditions because the sandstones contain many fossil marine shells and numerous (filled) burrows of mollusks. The gently folded strata are unconformably overlain by younger terrace deposits that correspond to the terrace deposits at Stop 7.

It is not difficult to deduce the sequence of geologic events that have taken place in the vicinity of Moss Beach, based on our observations at Stops 7 and 8. The earliest event was erosion of the granodiorite by wave action, yielding a planed off surface that is somewhat uneven. Next, over this surface, sand and gravel were deposited. Probably the planing process and deposition of sand and gravel took place simultaneously, perhaps under conditions similar to those prevailing near the shore now. The sand and gravel deposits were compacted and cemented to form firm rock. Then they were gently folded. Renewed erosion, in response to changes of land elevation with respect to sea level (lowering of the land or rise of sea level) allowed the planing-off process to occur again, followed by renewed deposition in which the material in the overlying terrace deposits was laid down. The final stage, which we observe at present, was preceded by a rise of the land relative to sea level, placing the surface of the upper terrace deposit 20 feet or more above sea level. The waves are presently eroding the rocks and producing a new wave-cut terrace.

Both local uplift and world-wide change in sea level probably were important in the sequence of events here. Also, it should be

**Figure 2-9.** Gently folded syncline exposed at low tide at Stop 8.

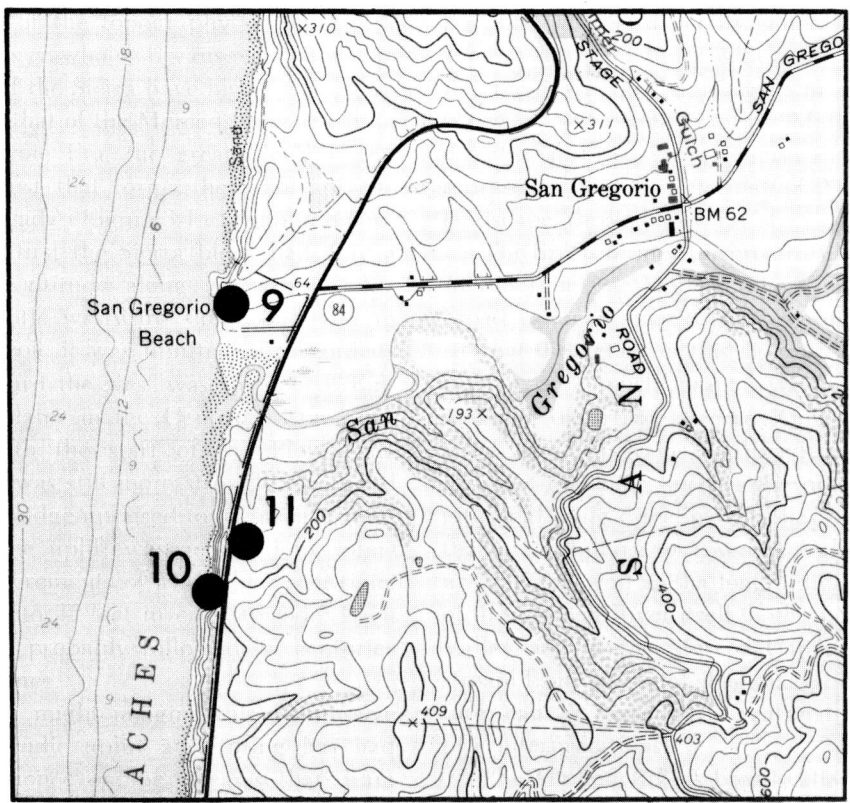

**Figure 2-10.** Map showing locations of Stops 9 to 11 in vicinity of San Gregorio Beach. East-west width of map is 1.7 miles. (From U.S.G.S. San Gregorio 7½′ topographic quadrangle.)

pointed out that it is a coincidence that the old wave-cut bench that levels the rocks exposed in the syncline (Figure 2-9) and vicinity is near present sea level. The old wave-cut bench is tilted, and is higher, for example, at Stop 7.

## SAN GREGORIO BEACH TO PIGEON POINT

### Stop 9: San Gregorio Beach (A)

San Gregorio Beach is part of San Mateo Beaches State Park. Figure 2-10 shows the locations of three stops in the vicinity. Stop 9 involves the vicinity of the parking area at San Gregorio Beach. Walk to the beach just below the parking area.

Interest here is centered on the exposures in the cliff. The parking area makes use of an uplifted marine terrace which is floored with terrace deposits that rest unconformably over gently dipping older strata (Figure 2-11). The terrace deposits consist of poorly sorted conglomerates, whereas the underlying strata are principally layered siltstones and fine sandstones

**Figure 2-11.** Cliff below parking area at San Gregorio Beach. Note terrace conglomerate unconformably overlying gently dipping Purisima strata.

which form part of the Purisima Formation. The Purisima Formation (named from Purisima Creek) is of Pliocene age. Its thickness is difficult to gauge, but it is on the order of several thousand feet. The Purisima has been broadly folded, as evidenced here by its gentle dip toward the northeast.

Both the terrace deposits and the exposed Purisima strata are of interest. The rounded cobblestones in the terrace deposits were probably brought by San Gregorio Creek, but the presence of the terrace gravels above the present level of the creek is explained by uplift of the land relative to sea level. The fact that San Gregorio Creek presently flows into the sea at sea level is explained by the fact that the creek has cut down through the deposits so that its gradient has been adjusted to the present relationship between the land and sea level.

The Purisima beds contain interesting sedimentary details. Fossils, consisting mostly of clam shells, are abundant in certain beds. Furthermore, many beds are intensely mottled as a result of the burrowing activities of animals when the sediments were still soft. The burrowing organisms, which probably included worms and clams, created passageways that probably filled in immediately afterward. The most spectacular display of burrowings is in the ceiling of a sea cave at the foot of the cliff facing the open sea (be careful, the cave is accessible only at low tide). Another sedimentary fea-

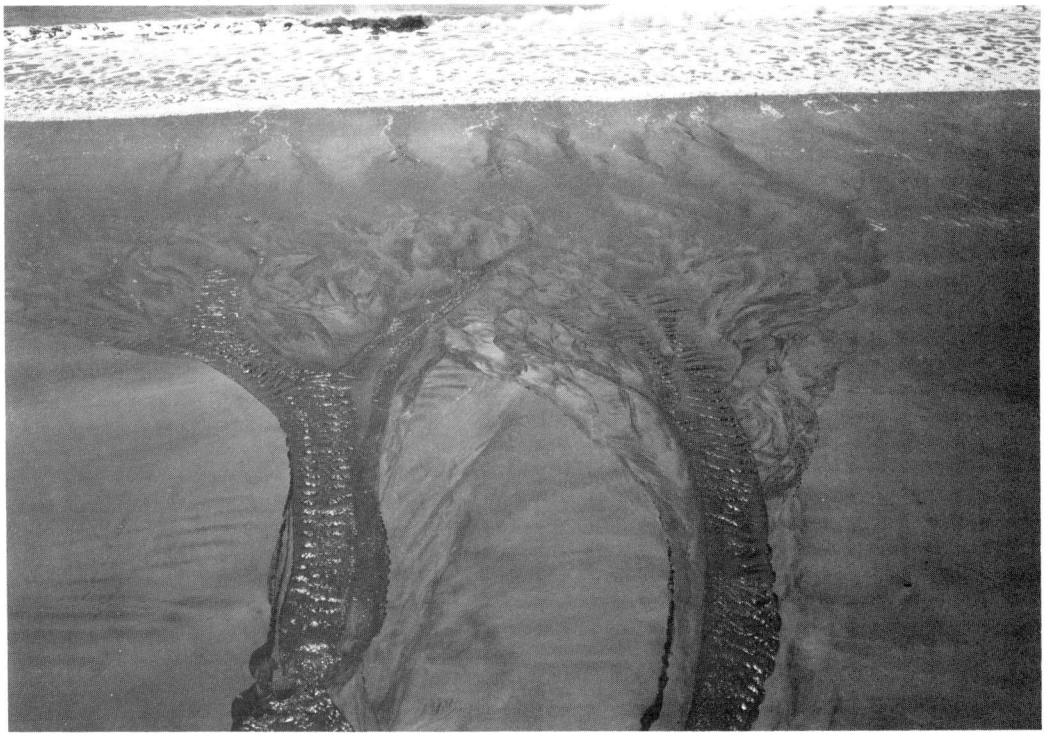

**Figure 2-12.** San Gregorio Creek entering sea at particular moment in time. Note intricate patterns of ripple marks in sand and ripples in water.

ture consists of the highly contorted laminae in some thick beds. The silty material forming the beds appears to have flowed like taffy while in an unconsolidated state.

The interplay between San Gregorio Creek and the sea provides opportunities to observe a sedimentation process. As the tide rises and falls twice a day, the sandy beach undergoes continual modification. San Gregorio Creek cuts through the beach deposits. Sometimes it carves out a channel with steep walls several feet high, exposing delicate layering in the sand. As the creek enters the sea it may create a small scale but spectacular braided distributary system (Figure 2-12) in which a variety of stream transport processes can be observed. While a small stream entering the sea is a common phenomenon, it is often an engaging spectacle. If you have a hand lens, note the appearance of the sand under low magnification. The sand here is finer than at Moss Beach. Why? There are a number of interesting minerals in the beach sands here and elsewhere in the Santa Cruz Mountain region. Consult the report by Hutton (1959) for additional information on beach sand mineralogy.

### Stop 10: San Gregorio Beach (B)

Stop 10 involves walking across the highway bridge over San Gregorio Creek (continue to park at the regular parking area), descending a short steep trail (watch for poison oak) and walking southward along the beach for roughly a third of a mile. We observe changes in the dip of the Purisima strata exposed in the cliff over this distance. As you proceed southward, you will note that the dip gradually decreases. The change in dip reflects the presence of a large anticline that plunges gently toward the east and the changes in dip that we observe mark the north flank and the crest of the anticline.

The top of the cliff is more or less horizontal. Just beneath the top of the cliff is a terrace deposit, which rests unconformably on the eroded edges of the Purisma beds. This terrace is obviously higher than that of Stop 9 and therefore is necessarily older. Note however, that the high level terrace deposits do not rest directly on the low level terrace deposits at Stop 9. If the higher deposits did rest directly on the lower level deposits (as at Moss Beach), the age relationships would be reversed, with younger above older.

## Stop 11: Upper Terrace Deposits

Stop 11 is conveniently reached by driving to roadcuts at the top of the hill immediately south of San Gregorio Beach. Here we can observe more closely the upper terrace deposits that we observed from a distance on the beach below. The deposits are mostly poorly consolidated clays and silts which tend to be eroded into miniature badlands in road cuts.

## Stop 12: Sand Dunes

Over much of the distance between Stops 11 and 12, we travel on the upper terrace which rests on the Purisima Formation. Traveling south as we approach Stop 12, however, the topography changes abruptly and we enter into a broad low valley. The valley is occupied by Pescadero and Butano Creeks which join shortly before they flow into the sea (Figure 2-13). The valley itself marks an important geologic change. Adjacent to the coast north of the valley, the Purisima Formation crops out, whereas south of the valley, much older strata (Cretaceous) crop out. The valley obscures the line where the two units come in contact with each other. The contact is probably a fault, although this is a debatable point and the geologic map (Figure 2-2) does not represent the contact as a fault. The topography is much different south of the contact, where high sea cliffs disappear, and instead, the Cretaceous rocks form a jagged rocky coastline (Figure 2-13) which contrasts with the smooth, straight beach and cliffs north of the contact, where the Purisima Formation crops out. Obviously the erosional resistance qualities of the Purisima Formation and the Cretaceous strata are quite different.

The location of the sand dunes at Stop 12 is indirectly influenced by the location of the boundary between the rock formations. The sand has been derived from the beach and swept along by winds, to be heaped into dunes in the broad opening afforded by the valley.

You can park adjacent to the dunes. If your visit does not coincide with high tide, it is worthwhile to walk northward along the beach from the sand dune area. Locally, in some of the recesses in the cliffs, the wind has heaped sand into small dunes. These small dunes are temporary, appearing and disappearing from time to time. Also, the cliffs provide excellent exposures of sedimentary details of the Purisima Formation.

### Stop 13: Pescadero Beach

Pescadero Beach is also a unit of San Mateo State Beaches State Park. Park your car opposite the junction of Highway 1 with the road that leads eastward to the village of Pescadero. The shoreline is jagged, with only small locally continuous stretches of beach. Our attention is drawn to the cliffs, consisting of steeply dipping Cretaceous strata that consist of sandstones and siltstones. Close inspection reveals a myriad of sedimentary details, such as small-scale cross stratification, contortion of beds due to flowage while still unconsolidated, and the presence of small fragments of carbonized wood.

Heavy surf is a major erosive influence here. Large swells and breaking waves impact steadily against the rocks projecting above the surf. We are observing, of course, the processes by which wave-cut terraces are formed. We can reasonably prophesy that if the elevation of the land remains static with respect to sea level for some time, that the submerged wave-cut terrace forming now will be progressively widened.

### Stop 14: Marsh

The valley containing Butano and Pescadero Creeks consists of a broad marshland that is immediately inland of an estuary or drowned valley. The marsh is slightly above sea level, and is an example of a sedimentary environment that we can observe more conveniently than most environments in which sediments are deposited. Plants are among the contributors to the marsh depos-

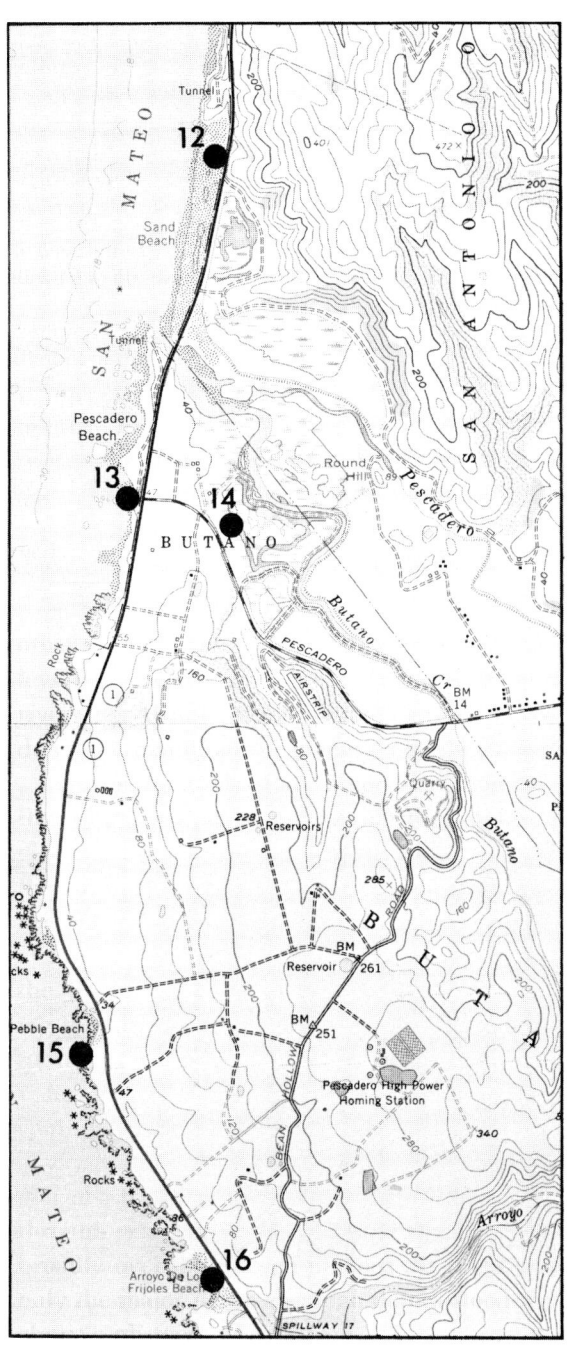

**Figure 2-13.** Location of Stops 12 to 16. East-west width of map is 1.5 miles. (From U.S.G.S. San Gregorio and Pigeon Point 7½′ topographic quadrangles.)

its. If you walk out into the marshland, you can feel the springy mass of sediment and compacted plant material. If this plant material were preserved from decay, deposits of peat would be formed. In fact, there is a parallel between the coal-forming swamps of the past and some of the peat-forming environments of the present. Marshlands occurring at or slightly above sea level provide a kind of peat-forming environment in many coastal areas of the world.

The estuary is now largely filled (Figure 2-13) by marsh material and stream sediment. The filling has taken place, as well as the development of the bar across its mouth (now largely covered by the sand dunes seen in the vicinity of Stop 12) after the development of the marine terrace that extends southward from San Gregorio Beach.

### Stop 15: Pebble Beach

Pebble Beach provides several interesting geologic observations. Walk to the beach via the narrow steps. True to its name, the beach is composed of small pebbles. When examined closely, particularly with a hand lens, the pebbles provide a beautiful sight. They are brightly colored, uniformly rounded, and highly polished. The pebbles appear to have been derived from chert beds in the Franciscan. We are immediately struck by the contrast with other beaches, which are sandy. Why the difference?

You may wish to inspect the strata that crop out at Pebble Beach (Figure 2-14). These rocks are readily observed by walking cautiously from the parking area toward the surf. Watch, however, for large waves that may sweep high upon the rocks. The strata, consists of steeply dipping, medium-coarse grained sandstone beds that contain lensing beds of conglomerate. The pebbles incorporated in the conglomerate, however, differ from loose pebbles on the beach. Note how the sandstones weather into intricate lacy patterns.

### Stop 16: Bean Hollow Beach

Bean Hollow Beach provides an interesting contrast with Pebble Beach. The beach material at Bean Hollow is sand rather than pebbles. It seems probable that differences in source material have had an important influence on beach composition. Apparently relatively few pebbles are supplied to Bean Hollow Beach. The sources of beach materials at Bean Hollow include not only those which are locally derived by erosion of rocks that crop out at or close to the beach, but also includes sediment brought to the beach by the stream named Arroyo de los Frijoles, which enters into the sea nearby.

### Stop 17: Pigeon Point

Pigeon Point is one of the highlights along Highway 1 between Santa Cruz and San Francisco. Pigeon Point is a popular scenic attraction, with its Coast Guard Station and classic white lighthouse standing on a projection of land that extends seaward. Pigeon Point is also famous for its tide pools, which contain a wealth of marine life, readily viewable at low tide. The complex ecology of the tide pools is easily damaged by man, particularly through the collection of specimens.

The principal geologic attraction at Pigeon Point is the conglomerate containing rounded boulders and cobbles. The conglomerate is overlain unconformably by relatively thin terrace deposits. The boul-

**Figure 2-14.** Cretaceous sandstones and interbedded conglomerate lenses at Pebble Beach (Stop 15).

ders and cobbles in the conglomerate are composed of a variety of lithologies, including granite, volcanic porphyry, and siliceous siltstone. The range of colors and sizes is impressive. The largest boulders are a foot or more in diameter. The boulders and cobbles are contained in a matrix of siltstone, somewhat like plums in a pudding. The proportion of boulders and cobbles (plums) to matrix (the rest of the pudding) is highly variable. In places, the boulders and cobbles predominate, whereas elsewhere the rock mass is mostly siltstone.

If you appreciate exotic landscape details, Pigeon Point's dissected, boulder-studded sea cliffs provide interesting scenery. Of equally great interest, however, are the scientific implications. What is the source of the admixture of cobblestones and boulders? The presence of volcanic prophyry specimens (containing crystals of feldspar that are conspicuously large in contrast to the fine-grained matrix) suggests an ultimate volcanic source, but the prophyry cobbles may have recycled. We can deduce, however, that there may have been outcrops of siltstone and of granite to supply the cobbles and boulders of those lithologies.

The heterogeneity of the bouldery deposits, including the erratic jumble of sizes, suggests that the deposits were deposited rapidly. In fact, we might describe them as being "dumped," much as a dump truck pouring out a mixture of ill-sorted debris. The dumping process, however, probably involved transportation of the boulders and cobbles (as well as finer material) by streams to the sea. The dumping process itself may have involved accumulation of the materials near the shore, accompanied possibly by periodic mass movement of the materials while in an unconsolidated state by submarine landslides. At best, we can only speculate on what happened, but these speculations may be reasonable in the light of our general knowledge concerning sedimentation processes that take place in the sea close to land areas that are rapidly eroding.

The sources of materials for the cobbles and boulders pose some interesting problems. For example, is there any direct evidence of the sources of the cobbles composed of volcanic porphyry? If we look landward for volcanics, we can find some in the vicinity of Mindego Hill, about 10 miles northeast, but these volcanics are younger (Miocene) than the Pigeon Point rocks (Cretaceous), and are of different composition, and therefore cannot be the source.

Our search is likely to be in vain if we consider that the present geography undoubtedly bears little relation to the ancient geography. Some of the geographic changes are related to displacement along the San Andreas fault, which may be as great as several hundred miles since Cretaceous time. If this is the case, the former sources of the boulders and cobbles may now be hundreds of miles distant, and we can do little more than speculate about their possible present locations, if indeed they are still in existence.

Careful scrutiny of the strata at Pigeon Point will reveal a few surprises. In places the strata are complexly folded, which is particularly obvious where they consist mostly of fine-grained siltstone. The folding probably occurred shortly after the beds had been deposited, while they were still in an incompletely consolidated state. This interpretation is in keeping with the general hypothesis that the deposits were "dumped" upon submarine slopes. Progressive slippage down the slopes probably caused the deposits to be deformed while they were still in a plastic state.

## DAVENPORT TO SANTA CRUZ

### Stop 18: Davenport

Davenport is a small coastal village whose principal industry is a cement plant. The site of the village occupies an uplifted marine terrace which is one of a series of uplifted terraces that characterize the local topography. The uplifted terrace is a remarkably plane surface which is overlain by a relatively thin terrace deposit (Figure 2-15). Bradley (1957) describes and interprets the origin of the uplifted marine terraces in the coastal areas of the Santa Cruz Mountains.

Our interest at Davenport is drawn to the rocks from which the terrace has been carved. Walk across the railroad track and down to the beach, following a steep trail. The strata exposed in the cliffs were probably deposited in the Pliocene and consist of siltstone that is distinctly bedded. If you pick up a piece, you may be surprised by its light weight. The lightness is due to the presence of diatomaceous earth. Diatoms are forms of microscopic one-celled plants that possess siliceous skeletons. These dia-

**Figure 2-15.** Sea cliffs carved in siltstone at Davenport (Stop 18). Note uniform, nearly horizontal surface of uplifted terrace. View looks northwest.

toms contributed large quantities of siliceous material to the rock and are partially responsible for its properties. The low density necessarily reflects the presence of a relatively high proportion of pore spaces to total volume of rock. The pores, however, are too small to be seen with the naked eye or even with a hand lens. Careful scrutiny with a hand lens may reveal the lacy, filigree forms of a few diatoms. Most of the diatom skeletons, however, have been dissolved, although they have contributed silica to the rock through reprecipitation of the silica that once formed their skeletons.

### Stop 19: Majors—Sandstone Dikes

Stop 19 is located at the intersection of Highway 1 and a side road extending toward the northeast (Figure 2-16). Watch for an elevated pipeline that extends over the side road and is plainly visible from Highway 1. The road cut immediately beyond the elevated pipeline exposes siltstone beds which have been intruded by sandstone dikes (Figure 2-17) that are impregnated with tar. The rock in the dikes weathers to a light gray or buff color, but when it is freshly exposed, the dark tar can be readily observed. The tar is the dried residue of oil in pores in the sandstone. We usually think of intrusive rocks as being igneous. Sandstone dikes, although rare, form an exception. It is clear that the dikes were forcibly intruded at Majors, wedging apart the siltstone in the process. The sand was probably soaked with fluid and behaved as a viscous liquid as it was intruded.

The origin of the tar sand is conjectural. It seems probable, however, that the sands moved up from beneath. Weathering and

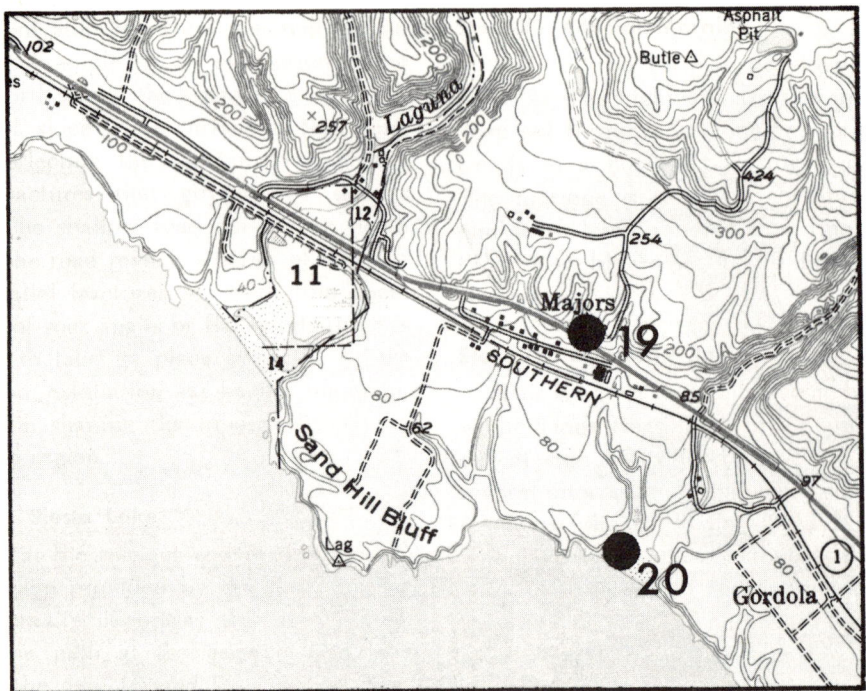

**Figure 2-16.** Stops 19 and 20. East-west length of map is 1.8 miles. (From U.S.G.S. Santa Cruz 7½' topographic quadrangle.)

erosion of the granitic rocks of the basement complex (which now lies at some depth beneath the siltstone) is a probable source for sand that was deposited before the siltstone was laid over it. Some of the unconsolidated sand, saturated with fluids, appears to have been forced upward and into the overlying siltstone. The siltstone, however, must have been consolidated when this happened, because it was fractured as a brittle solid, and the fluid-soaked sand moved along the fractures, wedging them apart to create dikes.

**Stop 20: Table Rock Beach—Sandstone Dikes**

Stop 20 is similar geologically to Stop 19 except that the sandstone dikes are much better exposed. Table Rock Beach is on private property and a small fee is charged for admission. At low tide, the sandstone dikes are readily observed, but at other times the cliffs in which the dikes are exposed are inaccessible.

**Stop 21: Natural Bridges Beach**

Stop 21 at Natural Bridges Beach State Park is an excellent place to observe the sculpturing effects of wave erosion. The natural bridges are arches formed by undercutting, and are being carved from a broad, nearly horizontal uplifted marine terrace. The arches occur in promontories projecting into the surf, which are attacked by waves on both sides. The siltstone, of which the terrace is formed, is sufficiently resistant to stand in vertical cliffs and to form the bridges.

## Stop 22: West Cliff Drive

West Cliff Drive in Santa Cruz follows along adjacent to the sea cliffs. Proceed directly east from Natural Bridges Beach State Park for about 1.25 miles, stopping where convenient. The scenery here (Figure 2-18) is similar to that at Stop 21, resulting from continuous wave attack of an uplifted marine terrace carved in the siltstone. A terrace deposit which forms a relatively thin veneer over the wave cut terrace, is being stripped away close to the sea cliffs. The newly exposed rock surface of the marine terrace reveals a pockmarked texture, which is a result of attack by rock-boring marine clams. This provides additional strong evidence that the rock surface on which the terrace deposits rest was formed below sea level.

The marine erosive processes acting on the fractured, brittle siltstone have produced a series of isolated marine stacks. The stacks are now surrounded by water, but were once continuous with the terrace, from which they have become progressively isolated as the material around them has been eroded away.

**Figure 2-17.** Intrusive sandstone dikes at Majors (Stop 19). Dashed lines have been added to photograph to outline some of larger dikes.

**Figure 2-18.** West Cliff Drive in Santa Cruz. Wave-cut terrace that has recently been exposed contains clam borings. Overlying deposits that have not yet been stripped away can be seen at upper left.

## Bibliography

Bradley, W. C. 1957. Origin of marine-terrace deposits in the Santa Cruz area, California. *Geol. Soc. America Bull.*, v. 68, no. 4, pp. 421-44.

Deals with uplifted marine terraces that flank the Santa Cruz Mountains.

Bramlette, M. N. 1946. *The Monterey formation of California and the origin of its siliceous rocks.* U. S. Geol. Survey Prof. Paper 212.

Deals with origin and characteristics of Monterey Formation, which is one of the formations of the area to which diatoms were important contributors. 57 pp.

Branner, J. C.; Newsome, F. S.; and Arnold, Ralph. 1909. *Description of the Santa Cruz quadrangle, California.* U. S. Geol. Survey, Geol. Atlas, Folio 163.

Report of pioneer geologic study in which the geology of part of the Santa Cruz Mountains was initially defined. 11 pp.

Cummings, J. C.; Touring, R. M.; and Brabb, E. E. 1962. *Geology of the northern Santa Cruz Mountains, California:* California Div. Mines and Geology Bull. 181, pp. 179-220.

Deals with structural, stratigraphic, and paleontologic aspects of Santa Cruz Mountains.

Hutton, C. O. 1959. Mineralogy of beach sands between Halfmoon and Monterey Bays, California. California Div. Mines Spec. Rept. 59.

Describes the heavy minerals that occur in beach sands of the Santa Cruz Mountains region. 32 pp.

# Chapter 3

# Sierra Nevada and Great Basin Provinces: Yosemite-Mono Craters Field Trip

**SIERRA NEVADA PROVINCE**

The Sierra Nevada Province is an elongate mountain range that is from 50 to 80 miles wide and more than 400 miles long. A particularly notable aspect of the Sierra Nevada province is its pronounced asymmetry. Ignoring local topographic details, the Sierra Nevada has a long, gentle western slope, but a precipitously steep eastern slope. The Sierra Nevada abuts the Mojave Desert at the Garlock Fault on the southeast and turns westward around the southern end of the Great Valley to merge into the Coast Ranges across the San Andreas fault on the southwest. The Great Basin lies to the east of the Sierra Nevada. The Sierra Nevada merges into the Cascade Range on the north.

The overall topographic shape of the Sierra Nevada—the gentle western slope and precipitous east-facing slope—is related to its gross structure. The Sierra Nevada has been carved from a huge block of the earth's crust that has been broken by faults along its eastern and southern margins and subsequently tilted westward (Figure 3-1).

The Sierra Nevada is famous for its scenery, particularly the "High Sierra" along the Sierra crest, which extends about 100 miles from Mt. Whitney, in the southeastern part of the range, to the vicinity of Tioga Pass at the eastern boundary of Yosemite National Park. The High Sierra is characterized by numerous lakes and a spectacular procession of 13,000 and 14,000-foot peaks that have been intensely sculptured by glacial erosion.

The topography and scenery of the Sierra Nevada are closely linked with the climatic influence of the range. Moist air from the Pacific moving eastward across California is elevated and cooled as it passes up the western slope and is frequently accompanied by precipitation of rain or snow. While the Coast Ranges have appreciable climatic influence, the higher elevations of the Sierra Nevada have even greater influence. On the east side of the Sierra, the descending air is warmed and can hold much more moisture than it contains. The Great Basin, to the east, is thus in the rain shadow of the Sierra. The Great Basin varies from

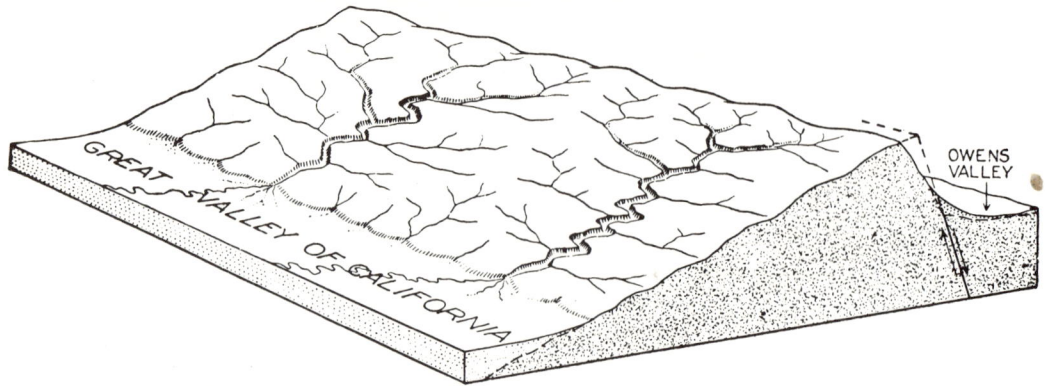

**Figure 3-1.** Idealized diagram showing tilting of Sierra Nevada block. That part of the Great Valley that lies adjacent to Sierra Nevada has been formed by burial of western part of Sierra block. (From Matthes, 1930.)

semi-arid in its higher elevations to extreme desert at lower elevations, as for example, in Death Valley. One of the effects of the rain shadow is that only a few of the higher peaks in the Great Basin have been sculptured by glaciers, whereas peaks at corresponding elevations in the Sierra have been intensely glaciated. The climatic influence of the Sierra is thus of long standing.

Most of the southeastern half of the Sierra Nevada and the eastern part of the northern half are composed of a complex of intrusive igneous rocks which are principally granitic. Collectively, these igneous rocks form part of the Sierra Nevada batholith, which is part of a still more extensive belt of intrusive igneous rocks that extends northward from the Peninsular Ranges in Baja California, through the Sierra Nevada and into western Nevada. These igneous intrusives were formed at various times during the Mesozoic Era. Radiometric age dates suggest that the oldest intrusive granitic rocks are roughly 200 million years old (Late Triassic or Early Jurassic) whereas the youngest intrusives in the core of the Sierra Nevada are about 80 to 90 million years old (Late Cretaceous). In the low western foothills of the Sierra, dark granitic rocks are Late Jurassic.

In the northern half of the Sierra Nevada, the batholith is flanked on the west by a belt of strongly deformed and metamorphosed sedimentary and volcanic rocks of Paleozoic and Mesozoic age. In the central and southern parts of the Sierra Nevada, scattered remnants of metamorphic rocks of these ages are found within the batholith, both in the western foothills of the Sierra, and along the crest of the range.

## YOSEMITE-MONO CRATERS FIELD TRIP ROUTE

The Yosemite-Mono Craters field trip route is shown in Figures 3-2 to 3-5. The first segment of the trip follows Highway 140, beginning in the foothills of the Sierras west of Mariposa, and thence along the Merced River canyon into Yosemite Valley. The second segment is along the Tioga Road portion of Highway 120, initially traversing the upland area north of Yosemite Valley, then across Tuolumne Meadows, and over Tioga Pass to the junction with Highway 395. Following a stop at Mono Lake, the route is along 395 briefly, and then east along 120 again, concluding at Mono Craters.

The field trip route crosses the Sierra Nevada province, and extends a slight distance into the Great Basin province. In addition, the Great Valley province is necessarily crossed approaching the Sierra Nevada from the west. While the Great Valley is of general interest to us, there are no field trip stops in it.

### Stop 1: Western Metamorphic Belt— Mariposa Slates

We can readily examine some of the metamorphic rocks along the western side of the Sierra Nevada batholith. These crop out in a broad belt known as the western metamorphic belt. Our route between Merced and Mariposa crosses part of the belt (Figures 3-2 and 3-3). The rocks include both sedimentary rocks and volcanics that have been weakly to moderately metamorphosed. The sedimentary rocks include dark siltstones that have been altered to slates, graywacke sandstones, and conglomerates. The volcanics include tuffs and breccias of andesitic composition, greenstones, and flows of basaltic lava. The sedimentary rocks are commonly interbedded with the volcanic rocks, and in places, intertongue with them.

Stop 1 provides an opportunity to observe dark siltstones and fine graywacke sandstones that have been metamorphosed to slate. These slates are commonly known as the Mariposa Slate of Jurassic age and represent part of the western metamorphic belt in which the rocks are of Triassic and Jurassic age. The stop itself is about six miles drive northeast of the village of Catheys Valley, along Highway 140. Outcrops of slate are prominent, particularly in the road cuts, so that the locality is easy to find. There is a slate quarry on the south side of the road, and a few hundred feet farther east, an unpaved road that extends northward joins the highway. It is convenient to park on the unpaved road because the shoulders of the highway are narrow.

The slates exposed at this stop (Figure 3-6) are dark graywacke siltstones and fine sandstones to which a moderate degree of metamorphism has imparted a distinct slaty cleavage. The metamorphism, however, has not obliterated features formed when the beds were initially deposited. Close inspection reveals that sedimentary features, such as fine cross laminae, are present in layers of siltstone and fine-grained sandstone.

### Stop 2: Western Metamorphic Belt— Calaveras Assemblage

This stop is about a 12-mile drive north of the town of Mariposa (11.6 miles from junction of Highway 49 and Highway 140). Look for a small bridge crossing Bear Creek and drive about 0.3 mile north of the bridge, parking on the east side of the road (this is the first good place to park north of the bridge). Bear Creek plunges over a fall held up by resistant greenstone, consisting of altered volcanic rock. The fall itself is difficult to walk to, but the greenstone is exposed in nearby roadcuts. The greenstone occurs in thick steeply dipping beds, and is part of the Calaveras assemblage. The Calaveras includes a variety of lithologies, and is Paleozoic in age.

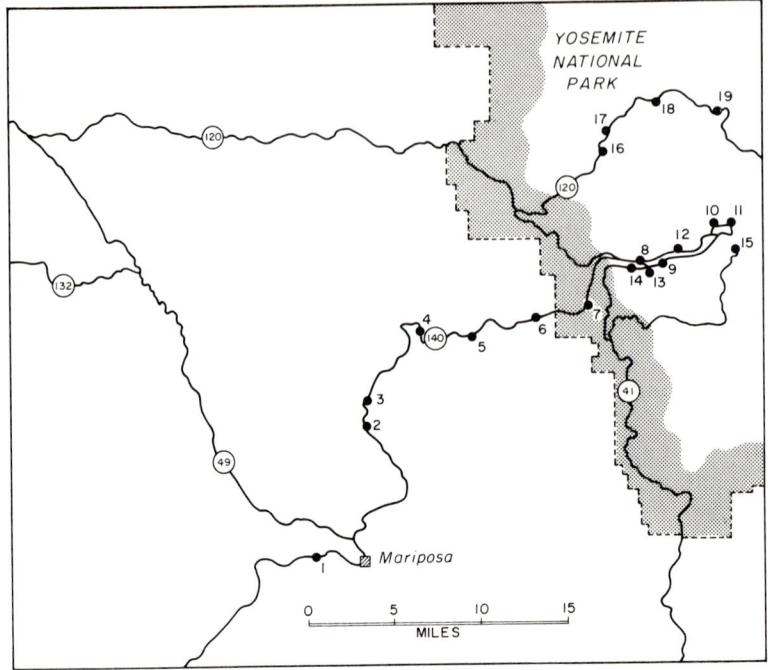

**Figure 3-2.** Map showing route and stops of first segment of Yosemite-Mono Craters field trip.

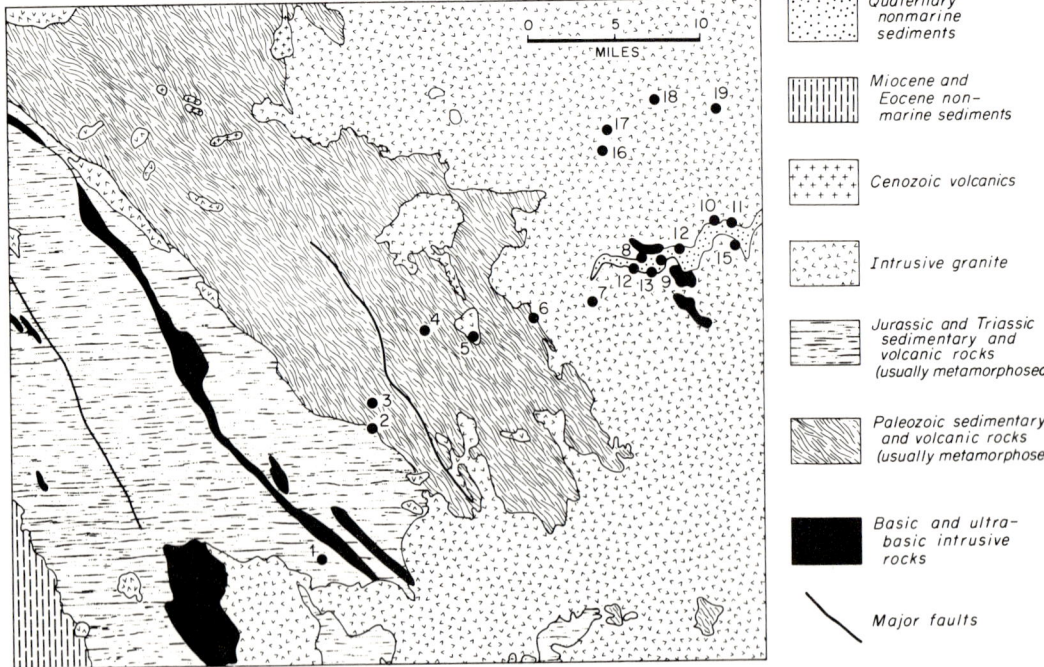

**Figure 3-3.** Simplified geologic map of area of Figure 3-2.

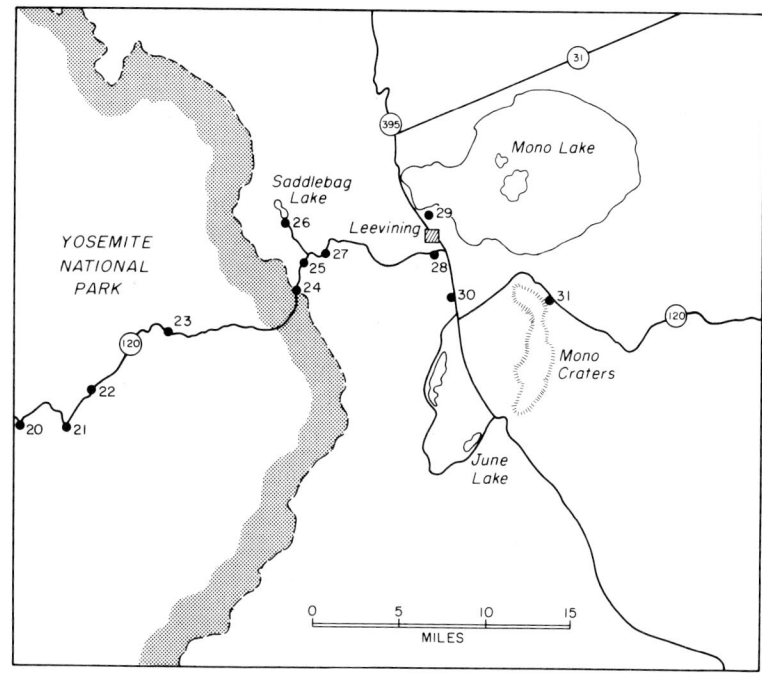

**Figure 3-4.** Map showing route and stops of second segment of Yosemite-Mono Craters field trip.

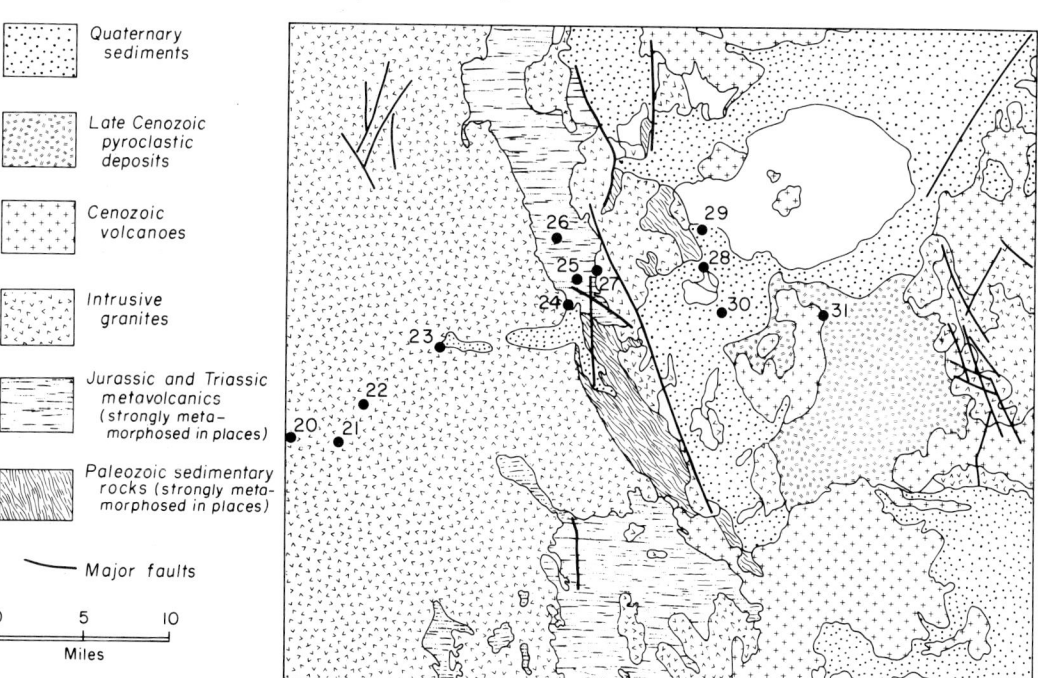

**Figure 3-5.** Simplified geologic map of Figure 3-4.

**Figure 3-6.** Slate outcrop along Highway 120 at Stop 1.

About 0.6 miles north (downhill) from this locality, you can observe the contact of greenstone with metasedimentary rocks, which consist mostly of dark siltstones, and are also part of the Calaveras. You may wish to stop, there being ample parking space on the east side of the highway. Notice that the rocks have well-developed cleavage which is generally parallel to bedding. In some places, however, the bedding and cleavage angles are different and therefore intersect each other.

## MERCED RIVER CANYON

### Stop 3: Glacial Outwash

The highway descends to the Merced River at Briceburg. Stop about 0.2 mile northeast of Briceburg. Here the Calaveras assemblage is well exposed (as it is for a number of miles along the Merced River), locally consisting of slate and chert. Our principal purpose in stopping here, however, is to examine gravel deposits that rest upon bedrock and may be observed in cuts along the south side of the highway. The gravels are glacial outwash consisting of a jumble of well rounded boulders of granite and granodiorite (Figure 3-7). Since the bedrock in the vicinity consists of slate and chert, the boulders must have been transported a number of miles by the river. It is very unlikely that the boulders were transported to this location by glacial ice, however, because the pebbles and boulders are quite well rounded, and the deposit is strati-

**Figure 3-7.** Glacial outwash gravels exposed on south side of Highway 140 about 0.2 of a mile northeast of Briceburg. Road cut is about 10 feet high.

fied. Moreover, the V-shaped profile of this part of the Merced River canyon strongly suggests that glaciers did not extend this far down the canyon. During the times that the upper parts of the Merced River canyon and its tributaries were occupied by glaciers, the latter supplied the Merced River with abundant rock debris which the river transported as glacial outwash.

**Stop 4: Folded Calaveras**

Stop at the geological exhibit which is about an 8-mile drive from Briceburg, and is marked by a prominent sign (entitled "Oldest Rocks of the Yosemite Region") on the west side of the highway (the Merced River flows north at this location). The Calaveras has been tightly and complexly folded, with steeply plunging fold axes. Although the folds can be observed immediately in back of and above the sign, it is worthwhile to view them adjacent to the river, particularly on the opposite side, where the polishing effect of the water has etched the thin, alternating layers of siltstone and chert into sharp relief. Probably the rock layers were folded very gradually at a time when they were several miles deep in the earth and were subjected to high confining pressure and elevated temperatures. The folding in the Calaveras is one of the effects of deep burial. Other influences include the intrusion of numerous granitic plutons, an example of which is provided

**Figure 3-8.** Merced River canyon in vicinity of Indian Flat. View is west-northwest.

by an isolated small pluton that occurs several miles to the east at Indian Flat, our next stop.

### Stop 5: Indian Flat

A small granitic pluton crops out in the vicinity of Forest Service Indian Flat Guard Station and campground. The rock is a coarse-grained gray granite which is generally deeply weathered. The contact of the pluton with the metamorphic rocks of the Calaveras Formation can be observed in several places. If you drive about a half mile east of Indian Flat campground, to White Wolf campground (which is on the south side of the road), you can discern the contact as you view the north side of the canyon. The contact dips steeply, with the lighter-appearing granite on the left (west) side of the contact, and darker metamorphic rocks on the right side.

While at this stop, it is expedient to consider the general form of the Merced River canyon. The profile of the canyon is distinctly V-shaped, with relatively smoothly sloping sides, and a narrow floor only a little wider than the river. Furthermore, the canyon winds and turns, changing directions relatively abruptly (Figure 3-8). The profile and plan of the canyon are a result of the downcutting effect of the Merced River. There is no suggestion that glacial ice has had a significant influence. Several miles to the east, as we approach El Portal, the canyon is both wider and less V-shaped, roughly marking the lower limit of glaciation in the Merced River canyon.

**Stop 6: El Portal**

Stop at the railroad exhibit on the north side of the highway. The Calaveras assemblage crops out in the vicinity, exposing layers of deformed marble containing non-precious garnets. Toward the east, the Calaveras assemblage is in contact with the Sierra Nevada batholith, which is our first view of this large batholith that forms a "sea of granite" about thirty miles wide, and extends along our field trip route from El Portal to the vicinity of Tioga Pass (Figures 3-3 and 3-5). The batholith contains a variety of granitic rocks and some darker rocks, such as diorite.

The Sierra Nevada batholith is thus a complex of individual igneous intrusions which were intruded at different times and with varying mineralogical and chemical compositions. The individual plutonic bodies that form the batholith presumably invaded and replaced the sedimentary and volcanic rocks that were previously present. The rocks of the western metamorphic belt are among the remnants of rocks that elsewhere served as "hosts" to the invading plutons. The abrupt contact between metamorphic rock and granite observed in the general vicinity of El Portal is significant because it is the interface between host and invader.

**Stop 7: Arch Rock Entrance Station**

As you cross the boundary into Yosemite National Park, the highway gradually steepens. Notice the large granite blocks in the floor of the canyon, some of them weighing hundreds of tons. Arch Rock Entrance Station receives its name from two huge blocks that lean against each other. The manner in which these blocks have been transported poses a problem. Although it might initially be supposed that they have been transported by glaciers, and left as a moraine, upon further consideration, it appears more probable that the blocks are simply talus blocks that have slid down from the cliffs above.

The canyon of the Merced River in the vicinity of Arch Rock is distinctly V-shaped, (it is known locally as Merced Gorge). The lack of a glacial U-shape is somewhat surprising, particularly when we see evidence of glaciation lower in the valley, in the form of the widened canyon in the vicinity of El Portal. Part of the explanation lies in the fact that there has been a succession of several distinct intervals of glaciation. The valley glaciers that reached El Portal and beyond were confined to earlier stages of glaciation. The glaciers that occupied the Merced River valley during the later glacial stages did not reach as far down the valley. The last one, for example, ended about a mile down valley from Bridalveil Fall. As a consequence, the present profile of the Merced Gorge reflects modification by running water, although its previous profiles probably reflected the effect of glaciation to a greater degree than at present. The topographic map (Figure 3-9) reveals the U-shape of Yosemite Valley above Merced Gorge.

YOSEMITE VALLEY

**Stop 8: Valley View**

Valley View provides an initial breathtaking glimpse of Yosemite Valley's size and grandeur. A better vantage point, which is north of the river, is not readily reached as

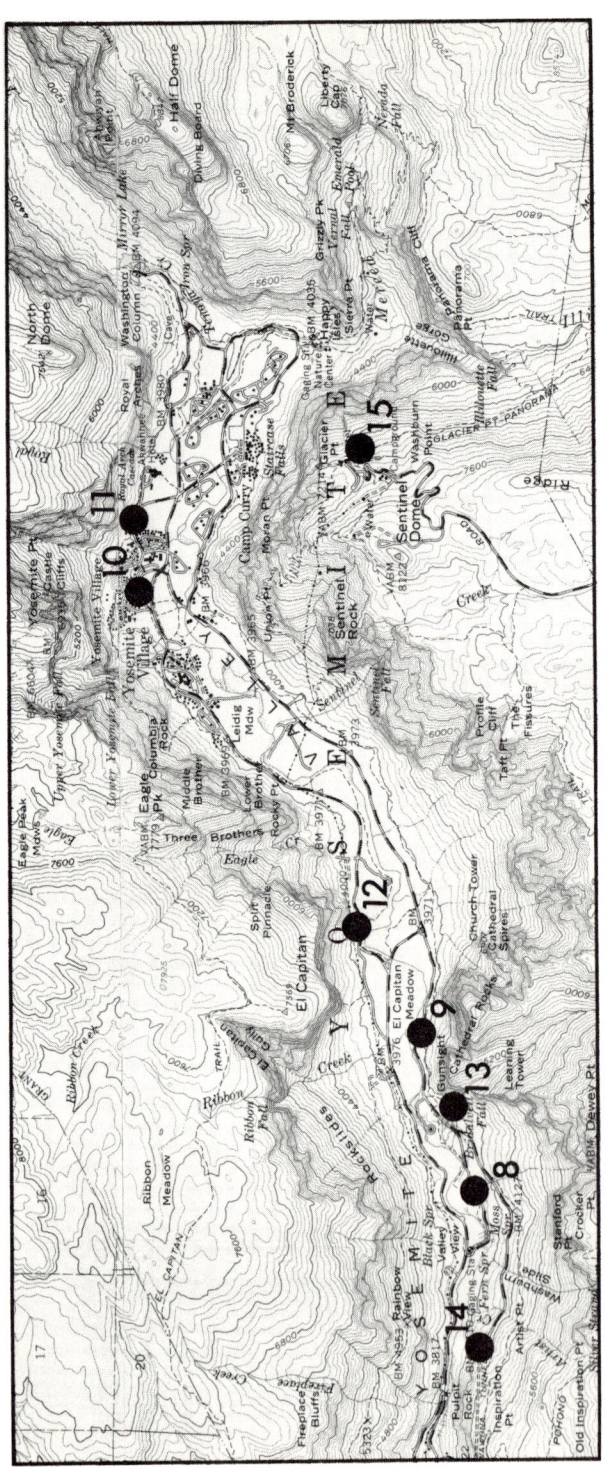

**Figure 3-9.** Topographic map of Yosemite Valley. Numbered field trip stops are shown. East-west length of map is 9.2 miles. (From U.S.G.S. Yosemite and Hetch Hetchy 15' quadrangles.)

you enter Yosemite Valley by car because of the system of one-way roads. As you drive eastward in the valley, you will be shunted to the south side of the valley floor. Nevertheless, an excellent view may be obtained by stopping at sign V-12 and walking across the meadow to the river. (The numbered roadside signs are keyed to the *Yosemite Road Guide*, which may be purchased at park headquarters.) Alternatively, you can postpone this stop until you drive westward out of the valley, stopping at sign V-11 along the one-way road on the north side of the valley.

At Valley View, you see El Capitan on the left (Figure 3-10) and Cathedral Rocks and Bridalveil Fall on the right. The U-shape of Yosemite Valley, in contrast to the V-shape of Merced Gorge, is well displayed here. The bottom of the U, however, is much flatter than is typical in glacially carved valleys. The explanation lies in the fact that the bedrock surface (the surface formed by hard rock) is almost 2,000 feet beneath the present floor of the valley between El Capitan and Cathedral Rocks. The present valley floor is a plain formed by partial filling of the valley by lake and stream sediments.

The sheer walls of El Capitan (which rise about 3,000 feet above the valley floor) and those of Cathedral Rocks reflect both

**Figure 3-10.** Yosemite Valley looking east from Valley View. El Capitan is on left, and Cathedral Rocks and Bridalveil Fall on right.

the massive nature (absence of joints or fractures) of the rocks of which they are composed, and the glacial erosive processes that shaped Yosemite Valley. Immediately north of Valley View, joints in the rock (diorite) are much more abundant, and as a result, the sloping cliffs are covered with talus blocks and the valley is wider. Between El Capitan and Cathedral Rocks, however, the valley is narrower and the cliffs much steeper. The relatively unjointed or unfractured rock that forms El Capitan and Cathedral Rocks has resisted erosion to a greater degree, which is reflected in the greater steepness of their cliffs.

Although a succession of glaciers shaped the Yosemite region, none of the glaciers in the Yosemite Valley completely covered El Capitan. The upper surface of the largest glacier reached only to the top of the steep face of El Capitan. On the other hand, the upper surface of this largest glacier was about 300 feet higher than the highest point of Cathedral Rocks.

### Stop 9: El Capitan Meadow

This stop (road sign V-13) provides an excellent face-on view of El Capitan. The rock exposed on El Capitan's cliffs is principally granite, but careful observation reveals that it is not a homogenous, uniform mass of granite, but instead, the rock face exposes dikes of quartz diorite that cut across the granite. On the opposite side of the valley, on the faces of Cathedral Rocks, light-colored dikes that are nearly horizontal intersect a complex of darker rocks.

As you drive farther east in the valley, additional excellent views of El Capitan are available, as for example, that from sign V-17 during the morning when the sun is in the east.

### Stop 10: Visitor Center

Ideally, your first stop in Yosemite Valley should be at the park's Visitor Center. The Visitor Center provides a focal point for your exploration of Yosemite, not only in terms of a source of advice and publications, but also in terms of its excellent museum. Here, you may wish to purchase the *Yosemite Road Guide* (by R. P. Ditton and D. E. Henry), topographic maps, and other literature such as Matthes and Fryxell's *The Incomparable Valley: A Geologic Interpretation of Yosemite.*

One of the important displays in the museum is the large topographic model of the Yosemite region, on which the various outcropping formations are shown with different colors. The model forms, in effect, a three-dimensional geologic map. The model emphasizes that the igneous rocks of Yosemite are not just a homogenous mass of granite, but instead are a complex of different igneous rocks that have widely varying compositions and were intruded at different times. While many of the details of the igneous geology are beyond our immediate interest, it is worthwhile to note some of the general differences. Table 3-1 summarizes the lithologic characteristics of the principal igneous rock types. Their geographic distribution is shown by the museum's model, as well as on the geologic map of Figure 3-11. Specimens of the various rock units including Taft Granite, El Capitan Granite, Sentinel Granodiorite, Half Dome Quartz Monzonite, and diorite are on display. These different specimens emphasize the fact that the various rock types do have noticeable differences when examined closely. Viewed from a distance, of course, the different rock types cannot be readily distinguished from each other except by differences in gross coloration, as for example, the dark gray cast of diorite as contrasted with the lighter gray of the granitic rocks.

The walls of Yosemite Valley and Merced Gorge reveal that there are at least seven large irregular intrusive bodies or plutons. Radiometric ages of these plutons indicate that the oldest and youngest differ in age by roughly 12 million years, and that the intrusions were emplaced about 84 to 96 million years ago, during Late Cretaceous time. These plutons, in turn, are cut by intrusive dikes and sheets of at least a half dozen other kinds of rock. If Yosemite Valley and its immediate vicinity provide a representative sample of the Sierra Nevada batholith as a whole, it would appear that the batholith was formed by scores or hundreds of distinct intrusions or plutons which were emplaced over a span of many million years.

Before you leave the Visitor Center's museum, observe the distribution of the Half Dome Granite and the Sentinel Granodiorite. Note also the relationships between the Taft Granite and the El Capitan Granite, and the wedge of diorite that locally separates them in the vicinity of El Capitan.

Yosemite's various igneous rocks have been classified into groups or series according to their geometrical and age relationships with respect to each other. The oldest series (Table 3-1 and Figure 3-11) is termed the Western Intrusive Series. While the time sequence in which the intrusions took place is not fully known, careful examination of cross-cutting relationships (i.e., where one intrusive mass cuts across another and therefore establishes its relative age as younger than the other) indicates that the Taft Granite is the youngest in this series. The intrusion of the Taft Granite was preceded by the El Capitan Granite, which, in turn, was preceded by both the granite of Arch Rock and by the diorite of the rockslides.

The minor intrusive bodies contain a variety of rock types whose age sequence is poorly known. The one that occupies the largest outcrop area is the Bridalveil Granite, whose properties have influenced the form of Bridalveil Fall, which we shall visit later.

Relationships in the Tuolumne Intrusive Series are relatively well known. Here the sequence of intrusions of the four formations is clearly shown by field relationships and confirmed by radiometric age dates. Of the four formations, only two crop out in Yosemite Valley (the Sentinel Granodiorite and Half Dome Quartz Monzonite), the other two (Cathedral Peak Granite and Johnson Granite Porphyry) are exposed farther east, toward Tuolumne Meadows. Interestingly, the outcrops of the four formations are very roughly concentric, with the youngest formation (Johnson Granite Porphyry) in the center. A discussion of the origin of the Sierra Nevada batholith as a whole is provided in the section following the description of Stop 28.

### Stop 11: Church Bowl

Church Bowl is about 0.4 of a mile east of the Visitor Center. The view across the valley provides a good view of Glacier Point. Note the contrast, in the cliff face west of Glacier Point, between the Half Dome Quartz Monzonite, which is relatively free from joints, below, and the Sentinel Granodiorite, above, which is cut by many joints. In the center of Yosemite Valley, just south of this point, the bedrock floor of the

TABLE 3-1. Principal igneous formations of the Yosemite region, listed in general order of increasing age from top of table downward. From Peck, Wahrhaftig, and Clark (1966).

| | Name | Lithology |
|---|---|---|
| **Tuolumne Intrusive Series** | Johnson Granite Porphyry | Light-gray, fine-grained quartz monzonite prophyry containing a little biotite but no hornblende. |
| | Cathedral Peak Granite | Light-gray quartz monzonite containing abundant large phenocrysts of K-feldspar in a medium-grained groundmass containing both biotite and hornblende. |
| | Half Dome Quartz Monzonite | Light- to medium-gray granodiorite and quartz monzonite containing well-formed crystals of biotite and hornblende. Includes non-porphyritic and porphyritic facies, the latter resembling the Cathedral Peak Granite but containing better formed biotite and hornblende. |
| | Sentinel Granodiorite | Medium-dark-gray granodiorite and quartz diorite, variable in color and texture; typically well foliated. |
| **Minor Intrusive Bodies** | Diorite of the "Map of North America" | Very-dark-gray diorite similar to the diorite of the rockslides but finer grained. |
| | Quartz-mica diorite | Medium-dark-gray, medium-fine grained quartz mica diorite. |
| | Bridalveil Granite | Medium-gray, fine-grained granodiorite; moderately abundant biotite gives the rock a "salt and pepper" appearance. |
| **Western Intrusive Series** | Leaning Tower Quartz Monzonite | Medium-gray, medium-grained granodiorite; clusters of biotite and hornblende give the rock a speckled appearance. |
| | Taft Granite | Very-light-gray, fine grained quartz monzonite, finer grained and more uniform than the El Capitan Granite. |
| | El Capitan Granite | Light-gray, medium-coarse-grained biotite quartz monzonite and granodiorite. Vaguely porphyritic in part. |
| | Granodiorite of the Gateway | Dark-gray, medium grained quartz diorite and granodiorite. |
| | Granite of Arch Rock | Medium-light-gray, medium monzonite and granodiorite; contain characteristic poikilitic K-feldspar grains. |
| | Diorite of the rockslides | Very-dark-gray, coarse- to medium-grained diorite, quartz diorite, and gabbro. Very variable. |

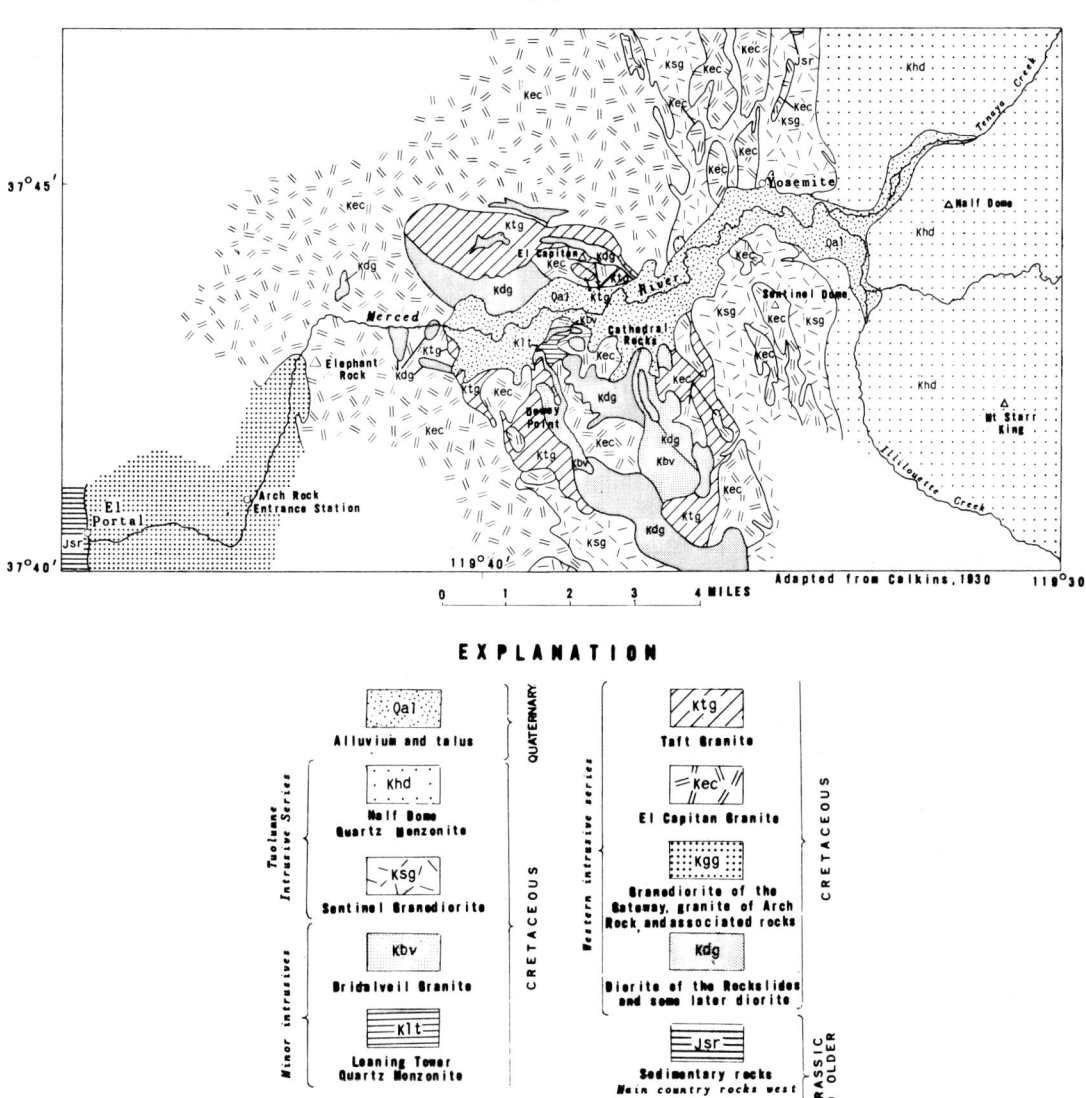

**Figure 3-11.** Geologic map of general area surrounding Yosemite Valley. Map emphasizes geographic distribution of different types of igneous rock. (From Peck, Wahrhaftig and Clark, 1966.)

valley is as much as 2,000 feet below the present valley floor surface, the valley having been locally filled to this depth with debris deposited by glacial streams and lakes.

Looking to the nearby cliffs on the north side of the valley, you may be able to discern (probably with some difficulty) dark-colored rock (Sentinel Granodiorite) which has been cut by gently dipping light-colored dikes (Half Dome Quartz Monzonite). This is one of the points of evidence that indicate the relative ages of the Sentinel Granodiorite with respect to the Half Dome Quartz Monzonite. Because the dikes (which extend from the Half Dome Quartz Monzonite) cut across the Sentinel, the Half Dome is necessarily younger. As the geologic map (Figure 3-11) indicates, we are close to the boundary between the Sentinel and Half Dome formations, the large mass of Half Dome Quartz Monzonite extending toward the east.

You probably will wish to visit some of the outstanding scenic attractions in the eastern part of Yosemite, including Vernal and Nevada Falls, and particularly Yosemite Falls. We will consider their geological significance, however, when we reach Glacier Point, from which we will have a commanding view of much of the Yosemite region.

## Stop 12: Foot of El Capitan

The view of El Capitan from its base is particularly impressive. Stop at road sign V-7. You may wish to walk along a primitive road from which a little used trail leads to the foot of El Capitan. If you miss the trail, you can readily walk cross country without a trail. As you survey El Capitan's face, notice the mottled effect. The dark surficial stains on El Capitan's face, however, should not be confused with dikes. A complex of intersecting, relatively thin dikes that appear lighter colored on weathering, is distinctly visible. Note the near absence of joints or fractures in the El Capitan Granite, a quality that has influenced the shape of El Capitan. The talus blocks that you see enroute to the foot of El Capitan include the medium-coarse-grained El Capitan Granite and the fine-grained Taft Granite, both of which are very light gray. Blocks of dark diorite dike material are also present.

As you look at El Capitan's face, notice the dark mass of diorite that forms a large "Map of North America" (Figure 3-12). This is a large irregular-shaped dike of diorite that has intruded the El Capitan Granite. The "Map" itself presents a remarkable resemblance to the map of North America—a wall map 2,000 feet in height, showing the continent all the way down to the Isthmus of Panama and a part of South America. Another large diorite intrusion farther east has an irregular shape that defies description. While the diorite that forms the "Map of North America" is younger relative to most of the other intrusive bodies, some of the thin light-colored dikes (which are pegmatite and aplite) appear to cut across the diorite intrusion and are, therefore still younger. Such are the complexities of igneous geology!

The valley floor contains a number of glacial moraines. If you walk about 200 feet west of sign V-7, you will see a hummocky moraine consisting of a jumble of angular granite blocks and finer debris which forms a low ridge. This is one of a series of terminal moraines left behind by receding glaciers. This moraine, and others served as temporary dams during part of the time when a lake (termed Lake Yosemite) was in existence. The lake was gradually filled by silt and sand by the Merced River and Tenaya Creek. Other moraines that are very slighty older and served as temporary dams may be observed at sign V-9. Lake Yosemite was about 5½ miles long in its maximum extent, its headwaters being near the base of Half Dome.

**Stop 13: Bridalveil Fall**

Stop in the parking lot that serves the trail to Bridalveil Fall and walk along the trail (about 1,000 feet) to the viewpoint at the base of the fall. The upper part of the cliff over which the fall plunges, consists of a thick, horizontal sheet of smooth-weathering Bridalveil Granite. Underneath is rough-weathering Leaning Tower Quartz Monzonite. These are among the minor intrusive bodies of the Yosemite Valley area. The Bridalveil Granite cuts across the Leaning Tower Quartz Monzonite, and is therefore younger than the Leaning Tower Quartz Monzonite.

**Stop 14: Tunnel View**

Stop at the parking lot near the east portal of Wawona Tunnel. Here, unfolded before you (Figure 3-13) is a magnificent panorama of Yosemite Valley—the great cliffs, Bridalveil Fall and its hanging valley,

**Figure 3-12.** Face of El Capitan. Note dark irregular dikes in center of cliff that forms an immense "Map of North America."

**Figure 3-13.** East-facing view of Yosemite Valley from Tunnel View.

and the U-shaped profile of Yosemite Valley. Observe the flat central floor of the valley forming the bottom of the U. Note also that Yosemite Valley narrows between El Capitan and the Cathedral Rocks.

The relationship of topography to rock type is especially important here. Observe the cliff on the north side of the valley, to the left of El Capitan. From a distance you can discern an intricate pattern of joints in the diorite (which is intersected by light-colored dikes). The fact that the diorite is so regularly and intensely jointed has allowed it to contribute abundant talus material that mantles the cliffs. El Capitan, by contrast, is composed of unjointed granite and stands in bold relief.

Before leaving the stop, examine the rock exposed immediately east of the entrance. Note the fractures or joints that dip at 20 or 30 degrees to the right and are roughly parallel to the slope of the land surface. Note also, the small masses of dark diorite that form inclusions in the lighter-colored granite (El Capitan Granite). The diorite inclusions may have been derived from the diorite of the rockslides.

The gently dipping fractures (or shells as they are sometimes called) which separate the rock into large slabs have played a major role in influencing the topography of the Yosemite region. Granite masses that otherwise lack regular joints tend to break along huge curved sheetlike fractures. In the process, the curved fractures or shells strongly influence the shape of the landscape. As overlapping shells are removed, rounded or domelike masses of rock are exposed, thus forming the many rock domes which are among the famous landmarks of Yosemite, including Half Dome. Thus, the domes are not shaped directly by glaciation, contrary to popular belief, although glaciers have doubtless been important in removing rock set free by the fracturing action at domes. The slopes in the area around Wawona Tunnel reflect the shells. Notice them as you drive toward Chinquapin, where the road to Glacier Point joins the main road.

### Stop 15: Glacier Point

There can be no doubt that the view from Glacier Point provides one of the most sublime spectacles of all in the Yosemite region. A walk of about three hundred yards from the parking area takes you to Glacier Point itself, whose elevation is 7,214 feet, or more than three thousand feet above the floor of Yosemite Valley. Toward the northwest, Yosemite Falls and the broad upland beyond are prominent. As your gaze swings from left to right, North Dome and the Royal Arches appear followed by Tenaya Canyon with Half Dome to its right (Figure 3-14). Finally, Little Yosemite Valley of the Merced River appears toward the

**Figure 3-14.** Tenaya Canyon viewed from Glacier Point. Half Dome is on right. Looking northeast.

**Figure 3-15.** Upper Merced Canyon looking east toward Vernal Fall (central foreground) and Nevada Fall (middle right), with Little Yosemite Valley beyond. Half Dome is on left. Succession of elevations of floor of Merced Canyon forms a giant stairway.

east, with Vernal and Nevada Falls in the middle foreground (Figure 3-15).

## TOPOGRAPHIC EVOLUTION OF YOSEMITE VALLEY AREA

Yosemite Valley has evolved under the combined influences of regional uplift, erosion by running water, glaciation, and alteration and decay of rock through weathering processes. Our attention is drawn to glaciation, however, as an exceptional influence in Yosemite's history. We have seen the canyon of the Merced River some miles downstream from El Portal, below the farthest known extent of the glaciers, where the valley's V-shaped cross-section is largely the result of river action and other ordinary erosive processes. It is reasonable to infer that the profile and cross-sectional shape of Yosemite Valley before it was glaciated was somewhat similar to this and to other valleys carved by running water. In other words, before Yosemite Valley acquired its U-shaped cross-section, it probably had a V-shaped cross-section. Before that, probably it was merely a broad, shallow valley prior to the uplifts which gave the Sierra Nevada its present height

To understand the geologic history of Yosemite Valley, we should consider the evolution of the topography of the Sierra Nevada as a whole, because many of the canyons of the Sierra have histories that are similar to that of the Merced River. Early in the Cenozoic Era, the great block of the earth's crust that forms the whole of the Sierra Nevada (Figure 3-1) began to be uplifted. At first the Sierra block had only a moderate altitude along its eastern edge and a gentle westward slope, but later in the Cenozoic it was uplifted again and its slope toward the west was steepened. Finally, still later in the Cenozoic Era, coinciding with the start of the Pleistocene Epoch, it was still more strongly tilted and its eastern edge upraised to its present great height. The uplift was achieved by intermittent increments of a few feet at a time, perhaps 1 to 25 feet in each episode. Each event was probably followed by centuries of quiescence.

As you might imagine this progressive but spasmodic uplift and tilting has had great influence on the streams that drain the Sierra block. Before uplift occurred, the main streams drained in diverse directions. When the regional slope became sufficiently steep, however, the main streams were realigned so that they flowed southwest, roughly parallel with each other and perpendicular to the long direction of the Sierra block. The present general course of the Merced River, along with other main streams, was thus established.

## Evolution of Yosemite Valley Before Glaciation

If we could step back several million years and view the Merced River in what is now Yosemite Valley, it would have had quite a different appearance. In fact, it is possible to reconstruct its general appearance at different times with assurance that the reconstruction is realistic in its gross features. The late Francois Matthes has done so, and the several figures that follow represent his interpretations of Yosemite's landscape through several stages.

The first is termed the broad-valley stage. At this time, the Merced River flowed in a broad valley (Figure 3-16A). We must realize that the details of this diagram are speculative—we must grant Matthes some measure of artistic license. On the other hand, these diagrams are based on various forms of evidence, such as the present longitudinal (long-direction) profiles of valleys that are tributary to Yosemite Valley (Figure 3-17) as well as the longitudinal profile of the Merced River itself, both in Yosemite Valley and in the Merced Canyon downstream from El Portal. The diagrams, therefore, can be regarded as reasonable scientific interpretations.

The Merced River's broad valley during the broad-valley stage formed in response to the tilt or gradient of the Sierra block at that time. The gradient was sufficient to allow the Merced River and its tributaries to erode a wide valley, above which the irregular uplands arose perhaps 500 to 1,000 feet over the valley floor. The tributary streams that joined the Merced River did so without passing over waterfalls or steep cascades.

The broad-valley stage came to an end when the Sierra block experienced renewed uplift and tilting, ushering in the mountain-valley stage (Figure 3-16B). There is debate as to when this occurred. Perhaps it was in the Pliocene Epoch, or possibly early in the Pleistocene Epoch. The effect of uplift and a steepened gradient for the Merced River

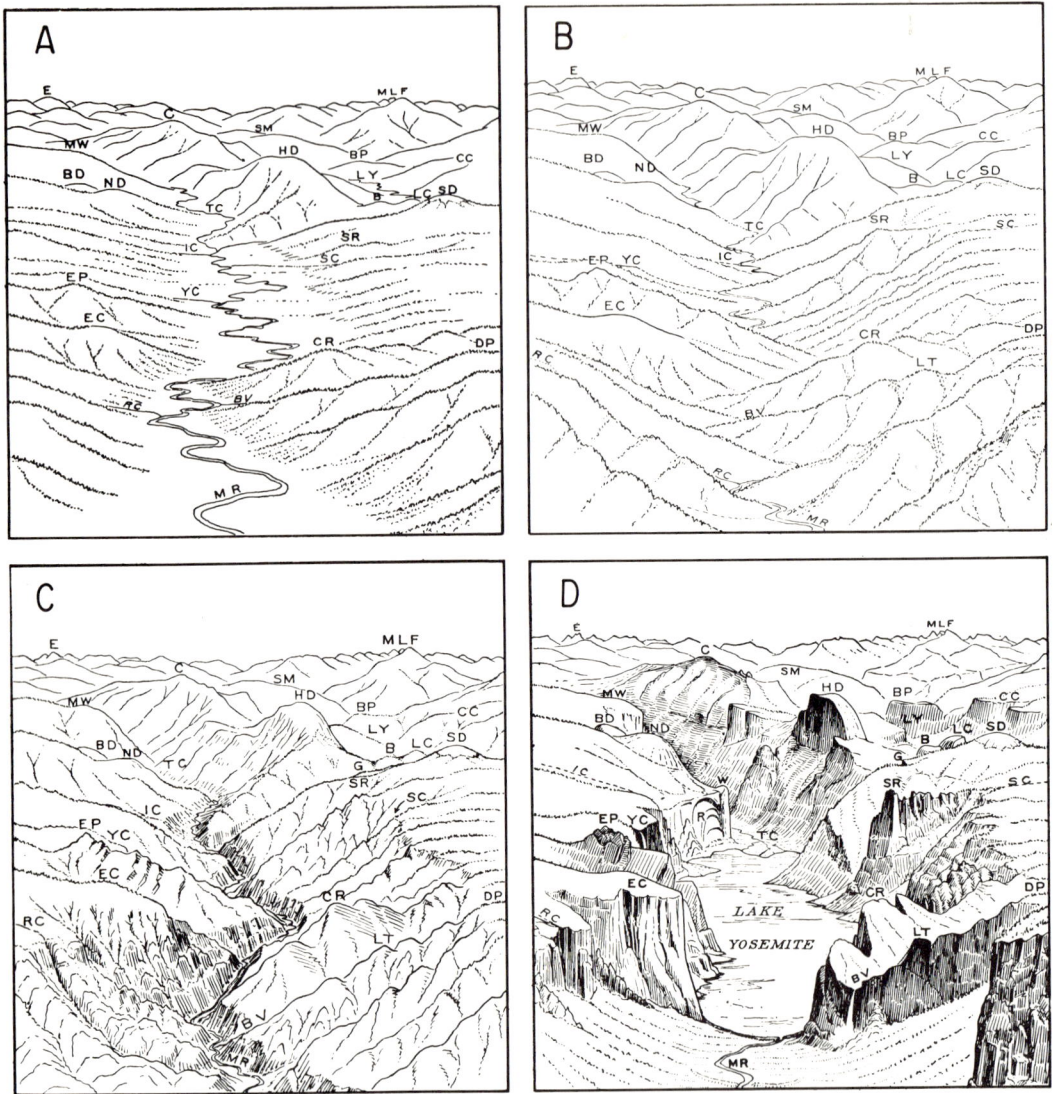

**Figure 3-16.** Interpretive bird's-eye views of Yosemite Valley as it might have appeared in successive stages of its development. (A) Broad-valley stage; (B) mountain-valley stage; (C) canyon stage; (D) immediately after last stage. The letters are keys to features that exist today, as identified below. (From Matthes, 1930.)

| | | | | | |
|---|---|---|---|---|---|
| RC, | Ribbon Creek. | E, | Echo Peak. | LC, | Liberty Cap. |
| EC, | El Capitan. | C, | Clouds Rest. | SD, | Sentinel Dome. |
| EP, | Eagle Peak. | SM, | Sunrise Mountain. | G, | Glacier Point. |
| YC, | Yosemite Creek. | HD, | Half Dome. | SR, | Sentinel Rock. |
| IC, | Indian Creek. | M, | Mount Maclure. | SC, | Sentinel Creek. |
| R, | Royal Arches. | L, | Mount Lyell. | CR, | Cathedral Rocks. |
| W, | Washington Column. | F, | Mount Florence. | BV, | Bridalveil Creek. |
| TC, | Tenaya Creek. | BP, | Bunnell Point. | LT, | Leaning Tower. |
| ND, | North Dome. | CC, | Cascade Cliffs. | DP, | Dewey Point. |
| BD, | Basket Dome. | LY, | Little Yosemite Valley. | MR, | Merced River. |
| MW, | Mount Watkins. | B, | Mount Broderick. | | |

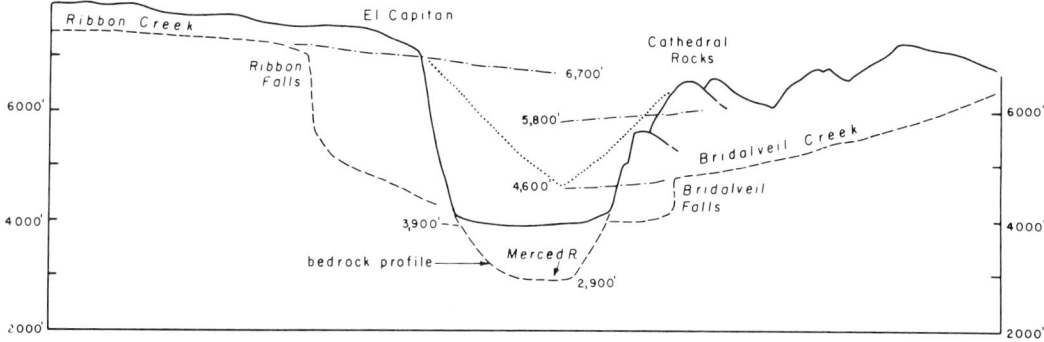

**Figure 3-17.** North-south cross-section across Yosemite Valley in general vicinity of Cathedral Rocks and El Capitan (solid line). Dashed lines are superimposed profiles of Ribbon Creek and Bridalveil Creek. Present profile of Ribbon Creek is extended into Yosemite Valley, representing broad-valley stage (line labeled 6700′). Profile during second stage (mountain-valley stage) is approximated by line labeled 5800′. Extension of Bridalveil Creek profile provides basis for approximation of third or canyon stage, whose profile is shown by lines of dots that form a V. Profile of bedrock below present valley is based on geophysical evidence. (From Wahrhaftig, Oakeshott, ed., 1962.)

was to cause the river to cut down, deepening the main valley as well as the lesser valleys of the tributary streams. Many of the tributary streams, however, no longer entered the main river on gentle gradients, but instead, descended steep cascades as they entered the main valley. The main valley was cut deeper and more rapidly than the tributary valleys, with the result that the tributary valleys were literally left "hanging" above the main valley.

Later, during the mountain-valley stage, there was additional uplifting of the Sierra block, accomplished by further steepening of trunk stream gradients. This caused the Merced River to erode its valley still deeper, so that it ultimately acquired truly canyonlike proportions, with a depth of 2,000 to 3,000 feet, ushering in the third or canyon stage (Figure 3-16C). The canyon (prior to invasion by glacial ice) was somewhat V-shaped in its overall cross section such as is the cross-section profile (Figure 3-8) of the Merced River at Indian Flat (Stop 5), for example. However, the river had carved a narrow inner gorge perhaps as much as 1,500 feet deep in places within the outer canyon, whose walls were less steep.

The narrowness of the canyon was accentuated by the numerous irregular topographic divides or spurs that projected into it. The location of spurs was inherited from the earlier stages, but the topographic relief was accentuated by downcutting of the trunk stream, and to a lesser extent by downcutting of tributary streams. The hanging tributary valleys were brought into still greater relief. The Yosemite region was then set for the drama of the ice age.

### GLACIAL HISTORY OF YOSEMITE REGION

The changes in climate at the outset of the Pleistocene Epoch influenced much of the world. Vast continental ice sheets spread over much of North America and Eurasia, and ice mantled the high mountains in many parts of the world. The succession of ice caps that mantled the Sierra Nevada was not, however, part of the continental ice sheets that spread over most of Canada and the northern United States east of the Rocky Mountains. The ice caps of the Sierra Nevada were distinct and wholly separate masses that originated locally, spreading over the higher elevations of the Sierra block, in a region extending from about 20 miles south of the latitude of Mt. Whitney, northwesterly to a point near the Middle Fork of the Feather River. At maximum extent, the Sierra glaciers collectively covered an area of more than 270 miles in length and up to 40 miles in width.

There are various forms of evidence indicating that multiple glaciation occurred in the Sierra Nevada. The evidence includes glacial erosional features of distinctly different degrees of preservation, suggesting that some are older than others. Such features include cirques, glacial stairways, U-shaped valleys, and glacial polish. Other evidence includes successive generations of deposits formed by or related to glaciers, such as glacial till and glacial outwash, as well as silt, sand and gravel deposited in glacial lakes.

It is possible to establish the relative ages of glacial deposits or glacial erosion features that were formed in different glacial stages. For example, glacial till deposits (which generally consist of an ill-sorted jumble of clay, silt, sand, pebbles, cobbles, and boulders) can commonly be ranked according to relative age by the state of preservation of the granitic material. Older tills have generally undergone a greater degree of weathering, and the granitic material that they contain tends to be deeply decayed; granite cobbles, for example, tend to be transformed to loose clay and sand. Furthermore, till deposits formed in earlier glacial stages are commonly more deeply dissected by erosion than younger deposits, unless the older deposits are overlain by younger deposits.

Four glacial stages are inferred to have taken place in the Yosemite region (Table 3-2). In addition, a still older stage, the McGee Stage, has been recognized on the eastern slope of the Sierra. Although there is no direct evidence of the McGee Stage in the Merced River drainage or elsewhere on the west slope of the Sierra, it is logical to infer that glaciers of the McGee Stage probably existed in the Yosemite area and elsewhere west of the Sierra crest.

The causes of the successive glaciation and accompanying world-wide climatic changes during the Pleistocene Epoch are still poorly understood. There can be no doubt, however, that the circumstances surrounding the accumulation of glacial ice in the higher elevations of the Sierra Nevada involved a surplus accumulation of snow over and above that lost through melting and evaporation. The Sierra receive heavy snowfalls today, owing to their great height and position in the path of the moisture-bearing winds that move across the land from the Pacific Ocean. The warmth during the summer, however, is sufficient to melt virtually all of the snow that falls, except for that which accumulates in some very small glaciers that still exist. If, however, the summers' warmth was insufficient to melt all of the snow that fell in the winters, the long-term effect would be gradual accumulation of ice as the residue of unmelted snow from each year was added to previous residues. Snow when compacted forms ice, so that continued accumulation of compacted snow forms glaciers.

TABLE 3-2. Glacial stages recognized in Yosemite Valley and vicinity.

| Name of Stage | Characteristics of features remaining from that stage | Maximum extent of ice down Merced River |
| --- | --- | --- |
| Tioga | Moraines fresh, undissected. Glacial polish locally preserved. Glacial lakes still preserved. | Head of Nevada Falls |
| Graveyard | Moraines moderately weathered; dissected by gullies 10 to 40 feet deep. Polish only rarely preserved. Lakes filled with sediment or drained. | Foot of Bridalveil Fall |
| Tahoe | Moraines weathered to 10 feet; dissected by gullies 10 to 70 feet deep. No polish remaining. No lakes remaining. | Few miles west of El Portal. |
| Sherwin | Moraines not preserved | |

Glaciers are able to slowly flow, even though ice is a crystalline solid. They are, therefore, not stationary and they move or flow with different velocities, depending on their mass, their cross-sectional area, and the steepness or gradient of the surfaces over which they flow. Although the down-valley terminus may become stationary, the ice continues to move slowly from the upper reaches down to the terminus, where it wastes away as fast as it arrives.

We can envision the initial gradual development of ice fields in the Sierra. At first there were probably only drifts or fields of compacted snow that survived from year to year in the heads of valleys that face the north or northeast. These grew sufficiently so that they began to flow, creating tongue-like ice streams or glaciers. Flowing ice, like flowing water, tends to seek the lowest elevation. As a consequence, the glaciers followed pre-existing stream valleys, joining other valley glaciers where the stream valleys join, eventually creating powerful ice rivers, or trunk glaciers that advanced down the main Sierra canyons.

During intervals of maximum glaciation, the snows were so abundant that the trunk glaciers, even though thousands of feet in thickness, could not carry the ice away from the gathering grounds in the High Sierra as fast as it accumulated. As a consequence, the branch glaciers overflowed, submerging the low divides between, and coalesced to form unbroken expanses of ice many miles across. These immense masses of ice continued to flow, of course, but instead of always following the paths provided by preglacial stream valleys, they also flowed through gaps and low ridges, and advanced broadly over the uneven uplands. At times, only the major peaks and high domes stood above a glistening sea of ice.

## Glacial Erosion of Yosemite Valley

The glaciers fed by the snow fields in the uplands to the east descended into Yosemite Valley via both Tenaya Canyon and the Merced Canyon. In all except the latest stage, the streams of ice coalesced where the two valleys join today. The resulting single stream of ice moved down Yosemite Valley, the position of its lower end varying from time to time. During the Tahoe stage (Table 3-2), the ice reached several miles west of El Portal, descending to altitudes as low as 1,700 feet above sea level. At its maximum, glacial ice completely filled Yosemite Valley and spilled over the rim on either side, covering, for example, Glacier Point and North Dome (Figure 3-18). During the Graveyard stage, the ice reached to the foot of Bridalveil Fall, whereas in the last major stage (Tioga stage), the ice did not reach Yosemite Valley, stopping just short of Nevada Fall.

The glaciers profoundly altered the earlier shape of the stream-carved Merced River canyon. The glaciers ground off projecting spurs, planed off parts of sloping canyon walls so that they became nearly vertical cliffs, and gouged a deep basin in the valley floor. The overall effect was to straighten and enlarge the valley, creating the U-shaped profile.

The glaciers eroded mainly by pulling loose and carrying away (plucking) blocks of rock defined by joints that intersect the rock. In addition, the glaciers ground and polished, although it appears that plucking was much more important than grinding and polishing as an erosive and valley-shaping process. The moving ice was most effective in deepening and widening the valley where the rocks are most closely jointed. Where joints are sparse, as at El Capitan, erosion was slower and as a result, the valley is narrower.

The proportionate amounts of erosion that had been accomplished by running water through canyon stage, as compared with that by glaciation, are shown in the

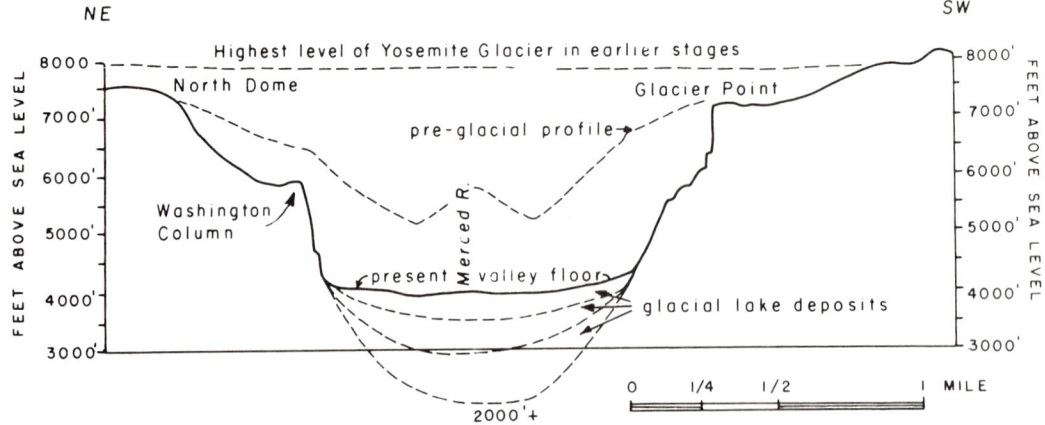

**Figure 3-18.** Profile across Yosemite Valley through North Dome and Glacier Point, showing present topography and inferred pre-glacial topography. (From Wahrhaftig, Oakeshott, ed., 1962).

two cross profiles (Figures 3-17 and 3-18). The glaciers cut even deeper, however, than the present valley floor. Geophysical studies involving the use of seismic waves indicate that the bedrock floor of the valley is generally more than 1,000 feet below the alluvial deposits that form the present valley floor. In the valley floor below Glacier Point, more than 2,000 feet of alluvium lie over bedrock in the center of the valley.

The earlier glacial stages left extensive systems of morainal ridges and isolated patches of till and erratic boulders on the uplands bordering the valley. These provide some of the evidence for the maximum geographic extent of glaciation. Moraines in the floor of Yosemite Valley are quite restricted, however, probably because some were washed away by swiftly flowing glacial meltwater, and partly because some were buried under lake sediments that now partly fill the valley. The moraines in the valley floor in the general vicinity of Bridalveil Fall and El Capitan may have served temporarily as a dam, helping to impound a lake (Lake Yosemite) that was subsequently filled with sediment. With the recession and gradual disappearance of the glaciers, the Yosemite area had assumed much of its present topographic form (Figure 3-16D). Lake Yosemite probably was filled with sediment during the later glacial stages. Postglacial processes, however, have influenced the topography in places. The rockslides or large talus deposits west of El Capitan have been created by disintegration of intensely jointed rock (diorite) both during and after glaciation.

The large rock domes, such as Half Dome, are among the most distinctive features of the Yosemite region. Contrary to earlier belief, these have not been created by glacial action. Instead, they result from the behavior of massive granitic rocks, such as the Half Dome Quartz Monzonite, which, spontaneously undergo fracture so as to create huge curving exfoliation shells of rock (Figure 3-21). The fractures may form as the granite mass which formed at great depth, expands after erosion has removed overlying material that was formerly several miles thick. The curving fractures allow shells of granite to separate and fall away, locally leaving dome-like features. While some granitic domes were covered by ice, others were not. Half Dome was never completely covered by ice. It owes its half-shape, however, to several factors. The general directional trend and the steep face of Half Dome (which slopes about 82° from the horizontal) was determined by a zone of nearly vertical joints extending in a northeasterly direction. The nearly vertical sheets of rock, defined by the parallel joints, were probably readily plucked away by Tenaya Glacier, which at one time reached within 500 feet of the top of Half Dome. However, the vertical sheets continue to fall away at the present time, without the help of a glacier. Probably there was never a former, more or less symmetrical exfoliation dome that was cut in half by glaciers moving down Tenaya Canyon. But the processes of exfoliation, coupled with the local joint system in the granite, have been the principal means by which Half Dome was shaped, and glaciation probably was of secondary importance. The presence of Tenaya Canyon immediately below the steep face of Half Dome, however, has no doubt been influential by providing a route through which exfoliation debris from Half Dome's steep face has been carried away.

The giant stairway that you observe in the Upper Merced Canyon where Vernal and Nevada Falls descend (Figure 3-15), impressive as it is, is only part of a much longer stairway that extends throughout the upper Merced Canyon, from Yosemite Valley to the foot of Mount Lyell—a stairway system about 21 miles long and rising 7,600 feet in that distance.

A staircase ascent by successive steps is a characteristic feature of many strongly glaciated canyons. While various hypotheses have been used in explanation, a very strong argument can be advanced that the succession of "stair treads" and "risers" in a glacial stairway is principally controlled by the intensity of jointing in the rock. Where the rock is cut by closely spaced, intersecting joints, glacial erosion is most effective because individual joint blocks are readily plucked away. By contrast, those parts of the body of rock that are sparsely jointed are less readily eroded because glacial plucking is relatively ineffective and abrasion is the principal process. Thus, both the depth of a glacially sculptured valley as well as its width is strongly influenced by the intensity of jointing in the bedrock.

Figure 3-19 is a schematic longitudinal section through a glacial stairway showing the mode of development by selective quarrying. Dash line AA represents the preglacial profile of the canyon floor carved by running water. The later profile BB illustrates the theory that the greatest depth of glacial erosion below the preglacial profile coincides with the greatest intensity of jointing in the bedrock.

## Tioga Road

### Stop 16: Pegmatite Dike

The next segment of the field trip involves travel over the Tioga Road, which is a segment of Highway 120. Drive about 8.9 miles northeast on Highway 120 from the highway junction at Crane Flat on Tioga Road. Watch the road cuts on the southeast side of the highway for the prominent pegmatite dike shown in Figure 3-20. You cross a small stream (the South Fork of the Tuolumne River) about 1.2 miles before you come to the pegmatite dike. If you overshoot, Smoky Jack campground is about 0.8 mile northeast of Stop 16.

We have seen pegmatite dikes high in the walls of Yosemite Valley. Here is a good opportunity to observe one at close range. The dike consists of an intergrowth of feldspar and quartz crystals. The boundaries of

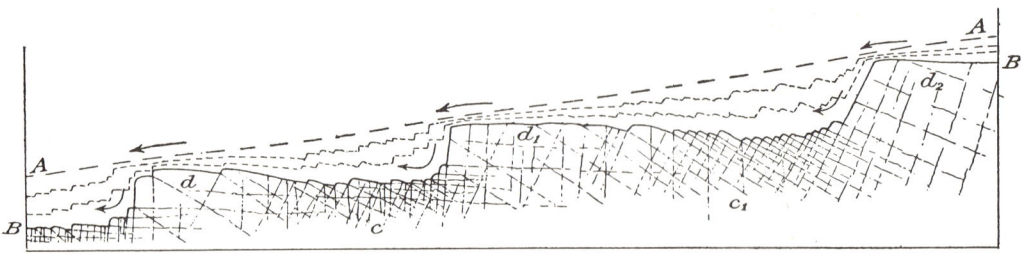

**Figure 3-19.** Longitudinal section of glacial stairway showing relationship between its profile and intensity of jointing. (From Matthes, 1930.)

**Figure 3-20.** Pegmatite dike exposed in road cut on southeast side of Tioga Road (Highway 120) at Stop 16. Also note numerous small inclusions of dark diorite in granitic host rock.

the dike are very sharply defined, there being an abrupt transition from the dike to the granitic rock into which the dike has intruded. The dike varies in thickness, and branches toward the northeast. While much of the dike material is relatively fine grained, toward the center of the dike the crystals of quartz and feldspar are very large, ranging up to several inches across. Also, in places there is a rhythmic alternation of more or less parallel plate-like masses of quartz and feldspar.

The granitic host rock contains a number of inclusions of dark diorite which range up to a foot or more in length. Although these inclusions are relatively irregular in shape, they are oriented so that their long dimensions are more or less vertical. We can speculate that they must have been formed earlier than the granitic rock (at least slightly earlier) and were subsequently incorporated in the intruding granitic rock. The pegmatite dike, on the other hand, is clearly younger than the granitic rock and the diorite inclusions because it cuts across both.

## Stop 17: Exfoliation Dome

Stop 17 is approximately 0.6 mile northeast of Smoky Jack campground (about 1.4

miles from Stop 16). Here you will see an excellent example of an exfoliation dome on the north side of the highway. The broad, bare rock slope is the surface of the dome itself, reflecting the configuration of the sheet fractures that govern the dome's shape. The shallow road cut on the north side of the road reveals a series of more or less parallel fractures. As each thin sheet or shell of rock spalls or flakes off, there is another to take its place beneath. As we have seen, exfoliation has been a major influence in shaping the topography of the Yosemite region.

### Stop 18: Siesta Lake

Stop at the turnout and read the interpretive sign provided by the Park Service (road sign T8 identifies this stop). Walk along the path at the edge of the lake toward the dam formed by a glacial moraine at the west end of the lake. The moraine dam is formed by a jumble of glacial debris, including boulders up to 10 feet in diameter. The fresh road cut on the north side of the highway opposite the turnout provides a good opportunity to observe the composition of the moraine. You will notice numerous other morainal deposits as you travel along the highway. Immediately east of Siesta Lake there are excellent examples of exfoliation domes, which in places are littered with glacial debris.

### Stop 19: Yosemite Creek View

Stop at road sign T13 or T14 and read the National Park Service's interpretive signs. The outstanding features here are the large exfoliation sheets. The road cut about 500 feet west of sign T13 reveals a succession of closely spaced exfoliation sheets (Figure 3-21).

### Stop 20: Glacial Moraine

Stop in the turnout provided at road sign T22. The highway passes by means of a deep cut through an elongate moraine that trends in a northeast-southwest direction. The moraine is composed of an irregular jumble of debris ranging from fine material to granite blocks up to 10 feet or more across.

### Stop 21: Olmstead Point

This is one of the major scenic highlights of the Tioga Road. Stop in the ample turnout at sign T24. Read the Park Service's several interpretive signs. Here you will get a commanding view of Tenaya Canyon, with Half Dome beyond toward the southwest, and Tenaya Lake to the northeast. Take the Py-We-Ack nature trail to the top of the nearby dome, where you will see glacially polished rocks with fine parallel scratches or striae formed by glacial ice as it passed over the rocks. Also note the excellent examples of exfoliation sheets on nearby domes.

### Stop 22: Tenaya Lake

Tenaya Lake is one of the scenic gems of the Yosemite region (Figure 3-22). Stop at the west end of the lake near the walk-in campground. Tenaya Lake occupies a topographic depression scoured out by glacial action. The domes that rise above the lake have been strongly influenced by exfoliation. If you walk a short distance immediately north of the highway, opposite the entrance to the walk-in campground, you will see fine examples of glacially polished surfaces that are littered with erratic blocks left by glaciers (Figure 3-23). Notice that the polished layer of the granite tends to "peel off" as thin flakes of rock. Because of

**Figure 3-21.** Series of closely spaced exfoliation shells at Yosemite Creek View.

this, polished surfaces are lost quickly, geologically speaking. Consequently, only the polish formed in the latest (Tioga) stage of glaciation remains today.

In many places, the exposed bedrock has been scoured by glaciers to form rounded, streamlined masses that range in size from a few tens of feet to hundreds of feet long. These are called sheep rocks (or more properly, the French term, "roches moutonées") in allusion to their resemblance to a flock of sheep in which the sheep stand close together so that only their backs are visible.

The granites near Tenaya Lake are of particular interest because they locally contain large isolated crystals of light-colored orthoclase feldspar up to four inches long and two inches wide. They are particularly well displayed on some of the glacially polished surfaces (Figure 3-23).

**Stop 23: Tuolumne Meadows**

Stop at road sign T29, where the interpretive signs describe the process by which Tuolumne Meadows have evolved. The Tuolumne Meadows area was once buried beneath a large ice field (known as the Tuolumne ice field) from which tongues of ice flowed down into Tenaya Canyon and the upper canyon of the Merced River, as well as elsewhere. The Tuolumne Meadows area

**Figure 3-22.** Tenaya Lake and glaciated mountains.

served as a gathering ground for glacial ice, and at times only the high distant peaks remained above the ice. The effect of this great mass of slowly flowing ice on Tuolumne Meadows itself was to scour the area. When the ice sheets melted away, a succession of shallow lakes remained. These lakes were subsequently filled with sand and gravel transported by streams, creating the broad meadows for which Tuolumne Meadows is named. Plants consisting of reeds and grasses also helped fill the lakes by furnishing organic material. Trees are presently colonizing some of the meadow areas and it is possible that the meadows will be eventually transformed into forest.

Note the glacially shaped and polished masses of rock, such as Pothole Dome, that rise above the meadows nearby.

**Stop 24: Tioga Pass Entrance Station**

The eastern boundary of Yosemite National Park and the crest of the Sierra Nevada roughly coincide with the contact between the Sierra Nevada batholith and masses of metamorphosed sedimentary and volcanic rocks that form an eastern metamorphic belt. The rocks in the eastern belt correspond roughly with those of the western metamorphic belt that we observed in Stops 1 to 6, although some of the rocks of the eastern belt are older. The metamorphic rocks of the eastern belt occur as partially isolated masses of metamorphic rock called roof pendants. The term roof pendant has been used to express the concept that the metamorphic rocks were left as masses suspended (like pendants) in the roof of the batholith.

We can appreciate the contrast between the igneous intrusive rocks and the metamorphic rocks, even from a distance. As you stop at the park entrance station, look toward the hill that parallels the highway, immediately to the west-northwest. The granitic masses are light gray, whereas the outcrops of metamorphic rock are brown.

If you are game for a walk of a mile or two, this general locality provides a good opportunity to observe the metamorphic rocks first hand. Take an unmarked trail that leads east from the entrance station toward Mt. Dana. You cross a half mile or more of Dana Meadow before reaching the slopes of Mt. Dana. Note the several glacial moraines that you cross in the meadow.

Most of Mt. Dana is composed of a succession of metamorphic rocks that are altered sedimentary and volcanic rocks originally laid down in the Pennsylvanian, Permian, and Jurassic Periods. The volcanics and sedimentary rocks are intimately

**Figure 3-23.** Glacial polish on roches moutonées north of Tenaya Lake. Note erratic boulders. Rock is Cathedral Peak Granite containing large individual orthoclase feldspar crystals up to four inches long.

intermixed. Lava flows and tuffs occur in the same sequence with limestones and graywacke sandstones and siltstones.

You will see the metamorphic equivalents of these rocks if you traverse in a north-south direction along the lower slopes along Mt. Dana's west side. Note that the fine sedimentary laminae in the siltstones and sandstones generally exhibit small-scale cross stratification that is brought into relief on weathered surfaces. The metamorphism has produced new minerals and rock types. Limestone has been transformed into marble, and some of the beds that were initially calcareous siltstone now contain abundant epidote in dull green, fibrous masses.

### Stop 25: Tioga Lake

Stop opposite the entrance to Tioga Lake campground (a U.S. Forest Service campground). Rocks exposed in the road cut on the west side of the highway are examples of dark siltstones that have been metamorphosed. The ridge rising immediately to the west exposes a complex of metamorphic rocks that weather dark brown, and an intrusive pluton that is light gray.

### Stop 26: Saddlebag Lake

Saddlebag Lake is another scenic gem of the Sierra Nevada. Take the 2½ mile long gravel road that turns off toward the northwest from Highway 120 at the junction about a mile northeast of Tioga Lake campground. Before you reach Saddlebag Lake on the side road, you will notice a spectacular view toward the west. In the distance you can see exposures of a large granitic pluton. The granite is intensely fractured and has been eroded by glacial ice that formed series of coalescing cirques.

Rocks in the foreground consist of brown-weathering metamorphic rock.

Saddlebag Lake is one of numerous lakes in the region formed by the scouring effect of glaciers. From the campground at the south end of the lake you can see a succession of glacially sculptured mountains toward the north. The sharp peak at the left end of the aggregation of distant peaks exposes a granitic intrusion, whereas the remainder of the rocks are metamorphic rocks.

### Stop 27: Ellery Lake

The lower part of Ellery Lake provides an excellent opportunity to examine the local relationships between a granitic intrusive and metamorphic rocks of the roof pendant. In the middle distance toward the southwest (Figure 3-24) there are steeply dipping metamorphic rocks that have been complexly folded. The steep cliff beyond exposes intensely jointed intrusives (quartz monzonite).

If you drive about ¼ of a mile past the dam of Ellery Lake, you can observe the contact between granitic rock and metamorphics in a roadcut on the northwest side of the highway. The relationships at the contact are complex. The metamorphic rocks to the right consist of steeply dipping altered siltstones. The contact between granitic rock and metamorphics is very irregular, with narrow dikes protruding from the granite into the metamorphics.

### Stop 28: Lee Vining Canyon

Highway 120 follows along the northern side of Lee Vining Canyon, providing a panoramic view of the canyon for most of its length. You may wish to stop at various places, the exact localities not being critical. The cross-sectional shape of the canyon is

**Figure 3-24.** View toward southwest from lower part of Ellery Lake, revealing folded metamorphics (dark) in foreground and quartz monzonite (light) in cliff beyond. Note network of joints or fractures in quartz monzonite.

that of a typical glaciated deep mountain valley, broadly V-shaped in the upper walls, but distinctly U-shaped in the lower part of the cross-section.

If we consider the entire length of Lee Vining Canyon, we can regard the canyon as the result of both erosional and depositional processes. The western part of the canyon's length represents extensive erosive sculpturing by glacial ice. The form of the eastern or lower part, however, is partly depositional in origin, for much of the material removed from the upper part has been deposited in the lower part as a series of glacial moraines.

Deposits of at least two glacial stages can be distinguished in Lee Vining Canyon. The oldest, widely exposed till in the canyon, that of the Tahoe stage (Table 3-2), forms the upper rim of the long lateral moraine that creates the south wall of the eastern half of the canyon's length (Figure 3-25). Inside and below this rim are extensive deposits of younger till (of the Tioga or youngest glacial stage).

We can envision the valley glaciers that formed these moraines. The glaciers received their "nourishment" of snow at higher elevations. They flowed downward, the lower end being defined at that point where the rate of flow and rate of melting were essentially in balance. The Tahoe stage glacier was larger, in both cross-section and length. Deposits of the Tahoe stage extend farther down the valley (to within a mile of the shore of Mono Lake) whereas those of the Tioga stage do not extend as far.

The details of the Tioga Till are best preserved. Where the Tioga Till is present, it tends to cover or obliterate older till deposits. The terminal end of the former Tioga glacier is marked by a succession of low crests or ridges (terminal moraines) in the floor of the valley. The oldest of these terminal moraines is the one near the mouth of the canyon. As the glacier gradually retreated, it left a succession of progressively younger terminal moraines similar to those shown schematically in Figure 3-26.

## THE SIERRA NEVADA BATHOLITH

Inasmuch as we have crossed the Sierra Nevada batholith in our travels from El Portal to Tioga Pass, it is worthwhile to consider some of the general aspects of the batholith and important problems of its origin. Much of the batholith is composed of quartz-bearing granitic rocks, but there are scattered masses of darker and older plutonic rocks, and remnants of metamorphosed sedimentary and volcanic rocks. The older plutonic rocks consist of relatively small bodies of diorite (as for example, the diorite of the rockslides of Yosemite Valley) and hornblende gabbro which occur as roof pendants within individual granitic plutons, or as septa (i.e., they serve as separators) between plutons. The metamorphic rocks also occur as roof pendants and as septa.

The granitic rocks which form the bulk of the batholith are in discrete plutons, which are either in sharp contact with each other, or are separated by septa of other rock types. Individual plutons vary greatly in size, their outcrop areas ranging from less than a square mile to more than 500 square miles. If the Sierra Nevada batholith is considered as a whole, it appears to consist of a few large plutons, and a great

many smaller ones which tend to be grouped between the large plutons.

Radiometric dates of various plutons in the central part of the Sierra Nevada batholith suggest that there are three widely separated epochs of plutonism, the first occurring roughly 190 to 210 million years ago, (during the Late Triassic), the second about 126 to 136 million years ago (Late Jurassic and Early Cretaceous), and the last about 80 to 90 million years ago (early Late Cretaceous). Presumably, the various plutons were emplaced during these epochs, but whether the intrusion processes were continuous or intermittent is difficult to determine.

The origin of the plutons is a problem of major importance. There is evidence that suggests that much of the granitic rock has crystallized from molten rock (instead of gradual transformation of previously existing rock in place as an extreme form of metamorphism). The evidence includes the sharp contacts of plutons with one another and with metamorphic rocks, and the presence of dikes and inclusions along contacts between plutons, from which the ages of plutons can be determined with consistent results. The plutons are pictured as having moved upward from a deeper source region because molten granitic magma has a lower density than solid rock of the same composition. As the plutons rose, the adjacent host rock is believed to have moved downward by gravity, thus providing room for the upward migration of the pluton. If this interpretation is correct, we can envision the process as one of exchanging molten magma for previously existing solid rock. Possibly some plutons may have been entirely detached from their source region at depth, but no field evidence for downward bottoming of plutons has been found.

The rising motion of plutons has had an effect on the host rocks into which the plutons were intruded. The rising magma squeezed the wall and roof rocks aside and upward. In the process, some host rock material was absorbed or was digested by the rising magma, but there is little evidence to

**Figure 3-25.** Panoramic View toward southwest, of lower part of Lee Vining Canyon. Opposite side of canyon is formed by lateral moraines deposited during last (Tioga) and earlier (Tahoe) glacial stages. Note terminal moraines in valley floor.

**Figure 3-26.** Idealized sketch of receding valley glacier which has deposited a succession of lateral moraines and looplike terminal moraines. (From Matthes, 1930.)

suggest that digestion was a major mechanism in emplacement of plutons, although locally it may have been important.

The overall structural or tectonic control of the Sierra batholith is a matter of guarded speculation. It seems reasonable, however, to assume that the granitic magmas were generated by melting of siliceous and aluminum-bearing components in pre-existing rocks in deeper regions at high temperatures. Laboratory experiments on the melting of granitic rock materials suggest that under pressure and in the presence of sufficient water, melting begins at temperatures between 600° and 700° C, and if other substances such as fluorine or chlorine are present, the temperature at which melting begins is even lower. The depth at which rocks of appropriate composition can be expected to melt to yield a granitic magma is difficult to specify because of inadequate knowledge about the generation and distribution of heat in the crust. In stable parts of the crust where heat is carried to the surface by conduction, a temperature of 600° to 700° C may be attained at depths of 20 to 30 miles. However, the melting could have occurred at greater depth; under the corresponding higher confining pressures, higher temperatures would be required.

The nature of the parent rock which melts to form granite is unknown. One possibility is deep burial of crustal rocks (perhaps sedimentary rocks) causing melting to occur. Alternatively, perhaps the granitic magma is formed by fractionation of more basic magma, or is formed by selective melting of certain components of basic or ultrabasic rocks.

The exposure of plutons at the earth's surface implies removal through erosion of great thicknesses of overlying rock. There is no simple way of estimating the thicknesses of eroded material, but perhaps it was on the order of 10 miles thick in the Sierra Nevada.

## GREAT BASIN PROVINCE

As we descend to Highway 395, we make the abrupt transition to the Great Basin Province. The most obvious difference is the climatic change—from the forested, relatively well-watered High Sierra to the barren desert or semi-desert. These climatic differences are of long standing, and they have had important influence on the topography and geology of the Great Basin.

In spite of the general aridity of the Great Basin, it too underwent drastic changes in climate during the Pleistocene Epoch. When the Sierra region was mantled with extensive glaciers, the Great Basin received much more moisture than today. While some of the high peaks of the Great Basin sustained mountain glaciers, most of the mountains in the province were not glaciated. An effect of the wetter, cooler climate, however, was to create a vast system of lakes in the valleys and topographic basins. Evidence for these lakes is overwhelming. Virtually all of them left prominent shore lines consisting of ancient beach deposits and shallowly indented wave-cut benches. The levels of these lakes fluctuated in response to long-term changes in precipitation. The present lakes and playas of the Great Basin represent the remnants of these Pleistocene lakes. Mono Lake is one of these lakes.

**Stop 29: Mono Lake**

Drive 1½ miles north of the town of Lee Vining on Highway 395 and then turn right on a gravel road that parallels the shore of the lake. Drive for ¾ of a mile or more on the gravel road, stopping where convenient. If you look back toward the highway, you can see several old shoreline levels (the horizontal topographic benches) along the mountain front. These mark some of the former levels of Mono Lake.

A low-altitude flight over Mono Lake would be particularly helpful in recognizing the succession of concentric shoreline features in the gently sloping lowland surrounding Mono Lake. Lacking an aerial view, however, we can examine some of these features at close range, as for example a sequence of ancient beach deposits of limestone. A series of freshwater springs supplies some of the water to Mono Lake. The spring water is particularly rich in dissolved calcium carbonate. When the spring water emerges in the lake, much of the calcium carbonate is precipitated as limestone.

The limestone beaches are one form of calcareous deposit. Another consists of the steep-sided white mounds of limestone formed at the springs themselves. Many of these mounds are now on land or project above the lake's present level. The mounds were formed when the lake was 15 to 20 feet or more higher than today. Forms of algae that secrete or incorporate calcium carbonate in their tissues were influential in creating the mounds.

**Stop 30: Bloody Canyon**

Bloody Canyon is the first canyon south of Lee Vining Canyon that has a long train of moraine deposits extending downward into the lowlands from the mountain front. Stop about three miles south of the intersection of Highways 120 and 395 at a convenient spot. The Sierra crest looms spectacularly, the deep indentation in the skyline being Mono Pass, which marks the upper source of Bloody Canyon's former glacier. As in Lee Vining Canyon and in the valleys of Grant Lake and June Lake to the south, several generations of glacial till deposits can be distinguished, including deposits of the Tioga and Tahoe stages, and till of the still older Sherwin stage.

**Stop 31: Mono Craters**

Turn east on Highway 120 at the junction with Highway 395, that lies about 5 miles south of Lee Vining Canyon. Continue in an easterly direction on 120 for about 6½ miles from the junction, where Highway 120 passes close to the north end of Mono Craters. Note the general form of the craters as you approach them, particularly the several circular domes that form the highest parts of the volcanic mass. Stop at that point where the edge of the craters is only a few hundred feet from the road. Notice that the area surrounding the Mono Craters is mantled with loose fine,

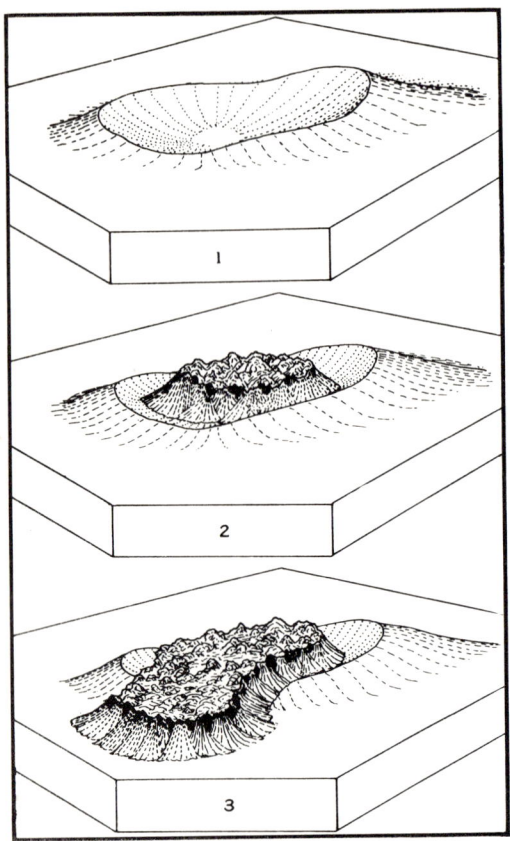

**Figure 3-27.** Schematic diagram illustrating three stages in development of individual domes and coulees at Mono Craters. From W. C. Putnam, (Reprinted from **The Geographical Review,** vol. 28, 1938.)

light gray pumice particles blown out of the craters by a series of explosions.

You may wish to walk up into the craters. This can be done relatively easily and safely. An abandoned mining road climbs up into an adjacent crater. In places the road is blocked, but you can walk around the obstacles easily. The surface of much of the Mono Craters is covered with blocky rubble of lava, much of it consisting of dark volcanic glass (obsidian), which in places displays fine flow laminae and contains layers of frothy pumice. Be careful to avoid being cut by sharp fragments of obsidian because they are, literally, fragments of broken glass.

The "craters" are not craters in the ordinary sense. Instead, the Mono volcanoes represent a relatively unusual type of volcanism which took place in a repetitious sequence of events in different parts of the extensive Mono Craters complex. Three main stages can be discerned. The first stage (inset 1 of Figure 3-27) involves development of explosion pits consisting of conical depressions whose sides slope inward. The material composing the walls and rim of the pits consists of pumice and fine volcanic fragments, although large fragments of obsidian are also abundant.

The second stage involved the rise of a stiff, cylindrical, essentially solid column of obsidian, which formed a viscous dome (rather than a fluid lava flow) in the floor of the explosion pit (inset 2). As the obsidian continued to ascend, it eventually spilled over the rim of the explosion crater, forming a coulee with a rough, blocky steep face (inset 3). The jagged cliff immediately before us at this stop is part of a coulee that extends over several square miles in the northeastern part of the Mono Craters complex.

The extreme ruggedness of the coulees is due to the fact that they hardened at the surface while the interior was still molten and continued to flow. The steep faces are due to the high viscosity of the lava, which is a consequence of its relatively high silica content. The present shape of a coulee probably differs very little from the shape at the moment that it "'froze."

The Mono Craters were formed very recently, geologically speaking. The gray pumice particles that were broadcast over the area surrounding the Mono Craters rest on moraines formed in the last glaciation, so that at least part of the volcanic activity that formed the Mono Craters is postglacial. In spite of the recent eruption, there is no sign of volcanic activity at Mono Craters today and we can only speculate whether there will be renewed volcanic activity.

## BIBLIOGRAPHY

BATEMAN, P. C. and WAHRHAFTIG, C. 1966. Geology of the Sierra Nevada, in *Geology of Northern California*, E. H. Bailey, ed. California Div. of Mines and Geol. Bull. 190. pp. 107-172.

> Masterful review of the geology of the central Sierra Nevada, providing both a general description of the geology as well as a discussion of the major problems. Contains extensive bibliography through 1965.

CLARK, L. D. 1964. *Stratigraphy and structure of part of the western Sierra Nevada metamorphic belt*. U.S. Geol. Survey Prof. Paper 410.

> Contains geologic map and description of metamorphic rocks exposed in Merced Canyon and elsewhere in the western Sierra metamorphic belt.

DITTON, R. P. and McHENRY, D. F., Yosemite Road Guide (may be purchased at Yosemite Park Museum).

> Useful guide to roadside observations in Yosemite National Park. The stops are keyed to the numbered roadside signs.

KISTLER, R. W. 1966. *Geologic map of the Mono Craters quadrangle, Mono and Tuolumne Counties, California:* U.S. Geol. Survey Map GQ-462.

> Detailed geologic map includes coverage of Mono Craters and Sierra Nevada crest just east of Tioga Pass.

MATTHES, F. E. 1930. *Geologic history of the Yosemite Valley*. U.S. Geol. Survey Prof. Paper 160.

> A classic geological report that provides the definitive description and interpretation of the Yosemite region. Although out of print it may be consulted in some libraries. 137 pp.

MATTHES, F. E. and FRYXELL, F. 1950. *The incomparable valley: a geologic interpretation of the Yosemite*. Berkeley: University of California Press.

> A superbly illustrated volume which has been written for the interested layman. May be purchased at the Yosemite Park Museum desk. 160 pp.

OAKESHOTT, G. B., ed. 1962. *Geologic guide to the Merced Canyon and Yosemite Valley, California*. California Div. of Mines and Geol. Bull. 182.

> Provides detailed field trip guide accompanied by description and interpretation of geology. 68 pp.

PECK, D. L.; WAHRHAFTIG, C.; and CLARK, L. D. 1966. Field trip guide to Yosemite Valley and Sierra Nevada batholith, in *Geology of Northern California*, E. H. Bailey, ed. California Div. of Mines and Geol. Bull. 190. pp. 487-502.

PUTNAM, W. C. 1938. The Mono Craters, California. *Geog. Review*, 28: 68-82.

> Description of geology and interpretation of origin of the Mono Craters.

# Chapter 4

# Cascade, Modoc Plateau, and Klamath Provinces: Lassen-Shasta Field Trip

## INTRODUCTION

Extreme northern California provides interesting contrasts in scenery and geology. These differences are reflected in the Cascade, Modoc Plateau, and Klamath provinces. The Lassen-Shasta field trip provides an opportunity to see representative portions of these provinces. Let us begin by reviewing some major distinguishing features of each province.

The most notable aspect of the Cascade province is a chain of young volcanic peaks that extends northward from Lassen Peak toward the Oregon border, and thence northward across Oregon and Washington, into British Columbia. The Modoc Plateau, which adjoins the Cascade province on the northeast, also includes young volcanic features, but these consist of smaller volcanic cones and other constructional volcanic features, as well as broad, basalt plains that are broken into large fault-blocks. It is the basalt plains that have given rise to the designation "plateau," although the Modoc region as a whole is not the monotonous elevated plain that the term plateau usually implies.

To the east, the Modoc Plateau grades into the Great Basin province. The boundary is somewhat arbitrary, for similar volcanic rocks occur in the fault-block ranges of the Great Basin as compared with those of the Modoc Plateau. The ranges of the Great Basin have greater relief, however, and the broad valleys between the ranges tend to contain thick stream-laid continental deposits and lake beds. Toward the south, the volcanic rocks of the Cascades and the Modoc Plateau overlap the metamorphic and plutonic rocks of the Sierra Nevada. Toward the west, the volcanic rocks of the Cascade Range overlap older rocks of the Klamath province.

These various provinces differ not only in the structure and lithology of the rocks, but also in the age of the rocks. Most of the exposed rocks of the Klamath province, as well as those of the Sierra Nevada province, tend to be substantially older than the rocks of the Cascade and Modoc Plateau provinces. Furthermore, the rocks of the Klamath and Sierra Nevada provinces have generally undergone much more structural deformation than those of the Cascade and Modoc Plateau provinces.

The route of the Lassen-Shasta field trip is shown in Figure 4-1. The sequence of stops as described here begins in Lassen Volcanic National Park and extends to Mt. Shasta, but of course the trip could be made in reverse order. The general geology of the region is shown in the geologic map of Figure 4-2. Let us now consider the Cascade province in greater detail, inasmuch as the Lassen-Shasta field trip begins in the Cascade province.

## CASCADE PROVINCE

The Cascade province in Oregon, and in northern California as far south as Mt. Shasta, may be generally separated into two major divisions, the Western Cascade Range toward the west, and the High Cascade Range on the east. Rocks of the Western Cascade Range consist of a thick sequence of layers that include lava flows and beds of pyroclastic debris. These strata, which have a pervasive greenish cast, were laid down during the Miocene Epoch, unconformably over Eocene and Upper Cretaceous sedimentary rocks. The total thickness of the Western Cascade strata is roughly 12,000 to 15,000 feet. South of Mt. Shasta, however, the rocks of the Western Cascade Range are not exposed and the twofold division of the Cascade Range does not apply.

We shall turn our attention to the High Cascade Range, for it contains the relatively young volcanic features that are of particular scenic and geologic interest. The High Cascades originated largely during the Pleistocene Epoch, although it is probable that the earliest High Cascade volcanics were formed in the latter part of the Pliocene Epoch. During their early history (late Pliocene and early Pleistocene), the High Cascades were characterized by the growth of a north-south chain of large volcanic cones built up by progressive outpouring of relatively fluid olivine basalt and basaltic andesite. Because of the low viscosity of the lava, the forms of these volcanoes can be likened to a swordsman's shield, being composed of gently sloping flows which dip away from the vents from which the lava emerged. Following this interval of eruptive activity, a series of eruptions of andesite built up large cones, including Brokeoff Cone in the southwestern part of Lassen Park, which has played an early role in the geological evolution of present-day Lassen Park. Subsequently, other types of eruptions occurred, continuing at intervals into recent centuries. These events included viscous extrusion of dacitic lava, and explosive eruptions producing cinder cones.

### Lassen Volcanic National Park

Lassen Park is an area of scenic contrast and exceptional beauty. Of California's five national parks, it shares the advantage, with Kings Canyon National Park, of being relatively uncrowded. As its name implies, Lassen Park displays geologic features that are predominantly volcanic in origin. Not only does this park possess a variety of volcanic features, but it also has the distinction of containing the most recently active volcano in the contiguous United States (i.e., the adjoining 48 states). We shall begin at Sulphur Works (Figure 4-3) and then travel in a general

# 86 • Cascade, Modoc Plateau, and Klamath Provinces

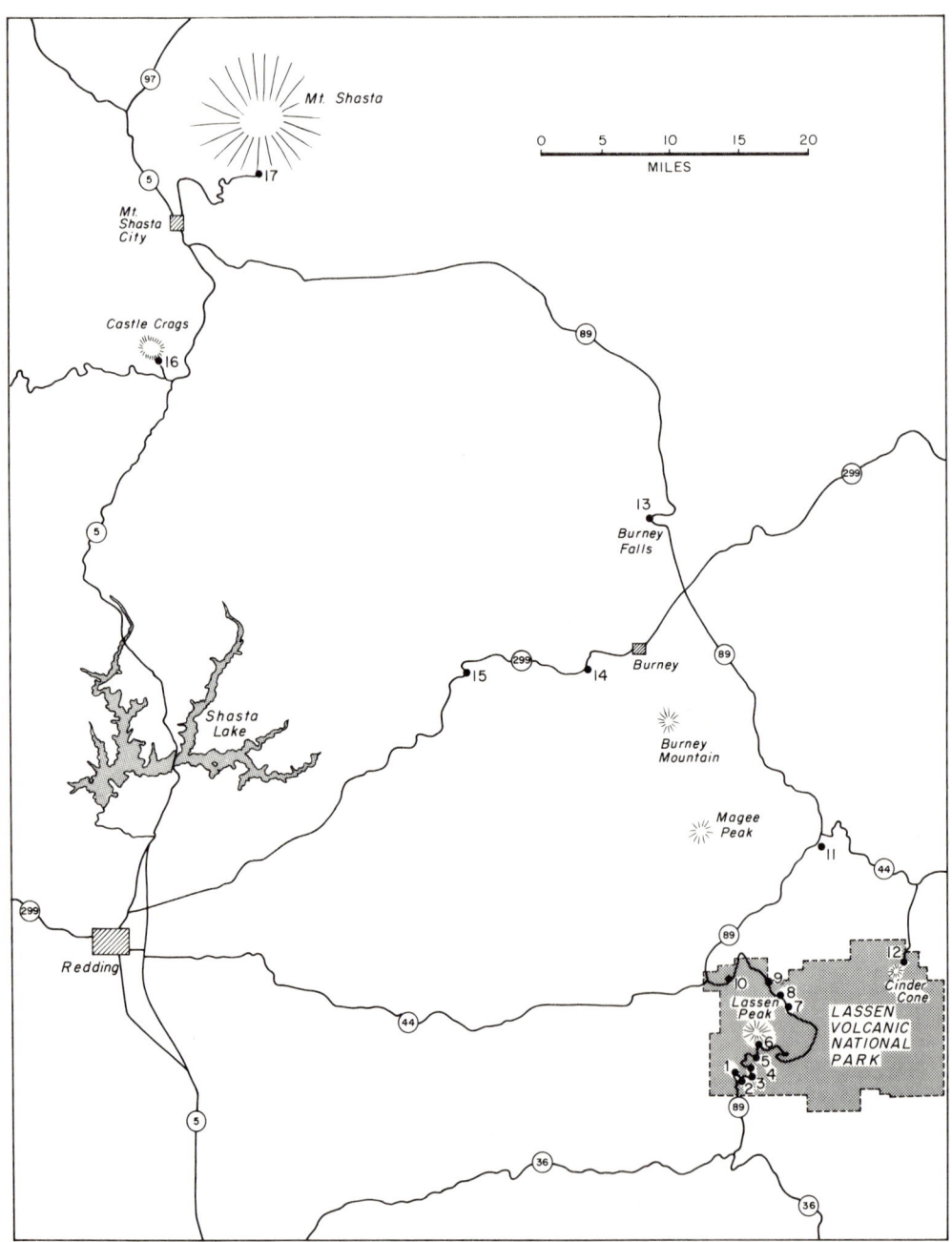

**Figure 4-1.** Map showing field trip stops and principal highways in region embracing Lassen National Park and Mt. Shasta.

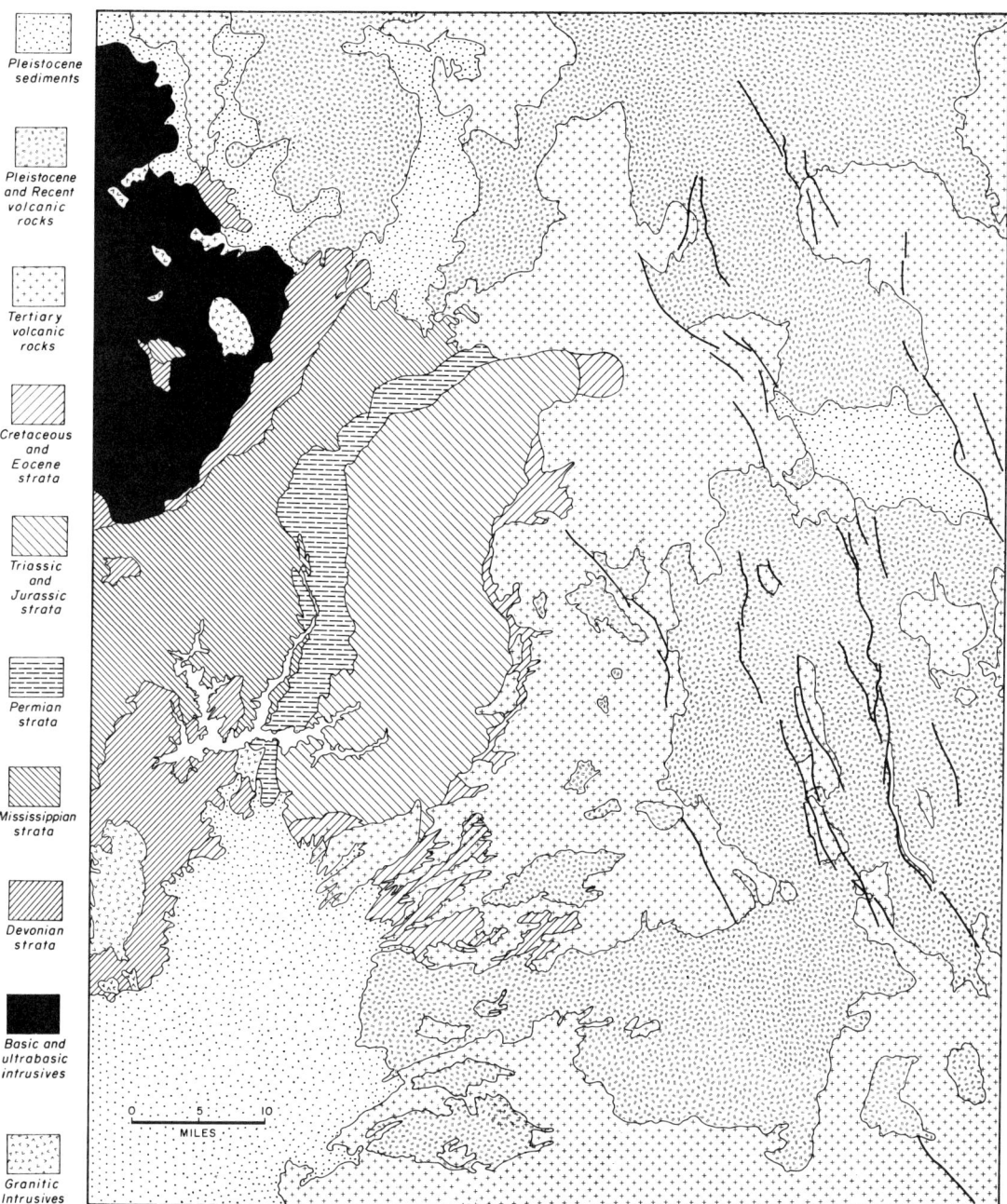

**Figure 4.2.** Simplified geologic map of Lassen-Shasta region. Area of map coincides with that of Figure 4-1.

northward direction on Highway 89, leaving the park immediately beyond Manzanita Lake, but returning via Highways 89 and 44, to Cinder Cone in the northeast corner of the park.

Before embarking on a tour of the park, you may wish to purchase the *Road Guide to Lassen Volcanic National Park*, which is available from the ranger at the entrance station. The *Road Guide* has been published in cooperation with the Park Service and is keyed to the small numbered roadside signs. We shall refer to these signs for convenience. Our stops are also shown on the topographic map of Figure 4-4.

**Stop 1: Sulphur Works**

Sulphur Works (sign number 5) is one of several places in the park where hot springs and fumaroles occur. The boiling springs, steam vents, and emanations containing sulphur are aspects of volcanism. A short, paved trail leads to the steam vents and springs. The odor of hydrogen sulphide (rotten egg odor) is very evident, and encrustations of native sulphur (small bright yellow crystals) occur around the steam vents and boiling springs. Some of the sulphur is in the form of sulphur dioxide which has been derived from oxidation of hydrogen sulphide. Sulphur dioxide reacts with water to form sulphurous acid, and subsequently to form sulphuric acid. Both acids are corrosive, so avoid getting the water and vapor on your skin and clothing (clothing tends to disintegrate when laundered after it has been exposed to acid).

The rocks in the vicinity of Sulphur Works are deeply decayed and have yellowish and buff tints. While some of the coloration is due to the presence of native sulphur, much of the coloration is due to the alteration of the rocks by the hot chemically reactive water, resulting in transformation of the original rock minerals to clays and iron oxides. Iron oxides, particularly limonite (hydrous iron oxide), commonly impart yellowish, brownish, or reddish tints to rocks.

**Stop 2: View of Brokeoff Mountain**

This stop (sign number 7) affords an excellent view of Brokeoff Mountain (Figure 4-5), which at elevation 9,235 feet, is the second highest peak in the park. Brokeoff Mountain consists of dipping layers of volcanic rock which are abruptly truncated or "broken off" on the north side, hence the name. Brokeoff Mountain is a remnant of a peak named Brokeoff Cone that existed hundreds of thousands of years ago and was subsequently destroyed by collapse. Brokeoff Mountain represents part of the south flank of Brokeoff Cone that was left after the central parts of the cone collapsed. If the dipping volcanic strata of Brokeoff Mountain are projected upward, they point toward the general vicinity of the center of Brokeoff Cone and provide a basis for estimating its former height, which was perhaps 1,000 feet higher than Lassen Peak's present elevation. The volcanic strata originally dipped away from the center of the cone.

Brokeoff Cone was a composite volcanic cone consisting of alternating lava flows and of beds of cinders and other volcanic debris blown out of the volcano. The collapse of Brokeoff Cone was probably caused by withdrawal of magma from beneath, the collapse taking place along a series of faults that intersected Brokeoff Cone. Mt. Conard, to the southeast, and the ridge between Brokeoff Mountain and Las-

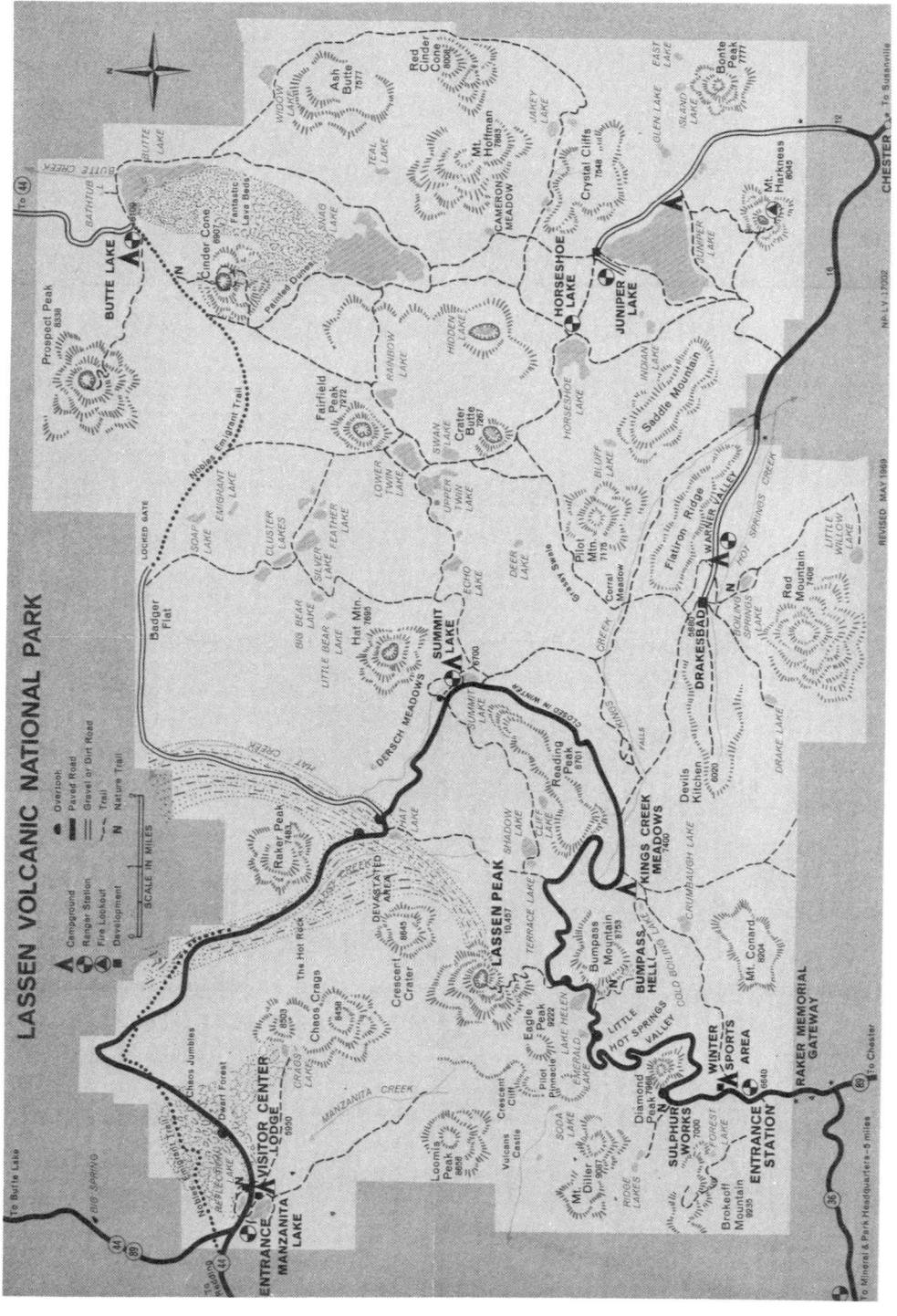

**Figure 4.3.** Maps showing principal geographic features in Lassen Park (From U.S. National Park Service.)

**Figure 4-4.** Topographic map of western part of Lassen Park showing location of field trip stops. (From U.S.G.S.)

**Figure 4-5.** Telephoto view of Brokeoff Mountain from Stop 2 (sign 7), showing distinctly layered volcanic rocks on north side of mountain. Rocks dip toward the south.

sen Peak, are remnants of Brokeoff Cone's former rim.

## Brokeoff Cone

Brokeoff Cone can be compared with other major calderas, such as Crater Lake in Oregon. A caldera is a large crater generally formed through collapse as a result of withdrawal of material beneath, although explosions may have been important in some calderas. For example, Crater Lake represents the collapsed summit of former Mount Mazama, whose slope and original height can be estimated by extending the present slopes of the cone-shaped mountain containing Crater Lake to the former apex of the cone. While Crater Lake's caldera is more or less circular, the caldera formed by the collapse of Brokeoff Cone is quite irregular, appearing to have involved a complex system of faults and other fractures whose exact locations cannot be determined readily. The fractures involved in its collapse, however, probably have served as conduits for movement of hot water. The extensive hydrothermal alteration of older lavas and the present location of hot springs and fumaroles, including Sulphur Works and Bumpass Hell, are probably closely related to the fault system of the old caldera. Figure 4-6 is a north-south geologic section, interpreted by Wil-

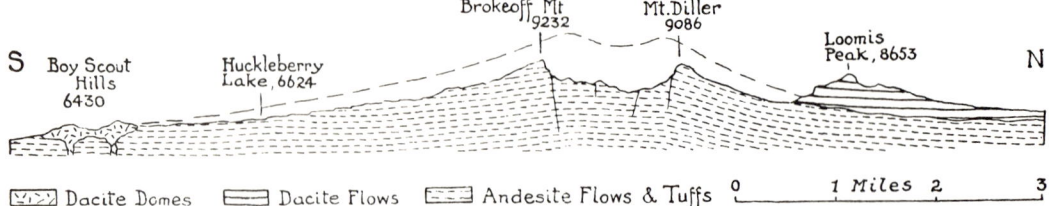

**Figure 4-6.** North-south geologic section near southwestern edge of Lassen Park, showing the former extent of Brokeoff Volcano, as interpreted by Williams, 1932a. (Originally published by the University of California Press; reprinted by permission of the Regents of the University of California.)

liams, which outlines the extent of Brokeoff Cone and shows its relationship to Brokeoff Mountain and to Mt. Diller and Loomis Peak. The center of Brokeoff Cone was probably close to Sulphur Works. During its early stages, the cone was built up by quiet outpourings of lava, forming a cone that was more or less symmetrical. The lava, being andesitic, was less fluid than basaltic lavas and formed steeper cones than Hawaiian-type shield volcanoes. Later in its development, eruptive activity at Brokeoff became more violent, with explosive outpouring of tuffs and breccias in addition to flows. Diamond Peak (viewed at Stop 3) is composed of tuffs, breccia, and flows formed during Brokeoff Cone's later stages.

**Stop 3: Diamond Point**

Diamond Point (sign 8) is on the southeastern flank of Diamond Peak. Diamond Peak was one of the conduits through which lavas moved upward during Brokeoff Cone's construction. Diamond Peak's summit is composed of volcanic breccia. The road cut at Diamond Point provides a good opportunity to view the breccia, which is interlayered with flows consisting of vesicular lava. Diamond Point also offers a good view of Mt. Conard, which is another remnant of Brokeoff Cone.

**Stop 4: Little Hot Springs Valley**

This stop coincides with signs 9 and 10, which are about a tenth of a mile apart. Stop 4 provides an excellent view of Lassen Peak from the south (Figure 4-7). Of particular interest here are the colorful altered lavas seen across the canyon of Little Hot Springs Valley. As at Sulphur Works, the alteration is the result of hot waters (and is thus appropriately termed hydrothermal alteration). The rocks, which were once hard, gray andesitic lavas, have been transformed by the acidic water into bright-colored clays. The reddish, tan, and purplish tints are principally due to the presence of iron oxides accompanying the clays. By driving to sign 10, a closer view of the altered lavas is available. These volcanic strata are also part of old Brokeoff Cone. The individual layers pinch and swell, and their susceptibility to alteration has varied greatly.

As you travel above Little Hot Springs Valley, notice its shape. In cross-section the valley is U-shaped, which is typical of a valley that has been modified by the erosive effects of glacial ice. By contrast, canyons in mountainous regions that have been carved largely by running water tend to be V-shaped.

### Stop 5: Bumpass Hell

Bumpass Hell lives up to its name, literally. It is an enjoyable kind of hell, however, with its sulphurous vapors, steaming hot springs, and bubbling mud pots. It is well worth visiting, being reached by a 1.3 mile (2.6 mile roundtrip) walk along the trail beginning at sign 17. Before embarking, pick up a trail-guide leaflet from the box near the start of the trail.

Bumpass Hell (Figure 4-8) extends over an area of about 16 acres or roughly an area 500 by 1,400 feet in extent, containing many steam vents and boiling springs. The hottest springs at Bumpass Hell are in excess of 200° F. At Bumpass Hell, water boils at 198° (instead of 212°) because of the elevation; the higher the elevation, the lower the atmospheric pressure. The fact that the temperatures of some of the springs are slightly higher than the boiling temperature is due to the presence of superheated steam. Superheated steam, as its name implies, is steam that is hotter than the boiling temperature that corresponds to its particular pressure. The heat is volcanic heat, of course, and we are reminded that the hot spring and fumarolic activity at Bumpass

**Figure 4-7.** Lassen Peak view at stop 4 (sign 9), located on west side of Little Hot Springs Valley. Distinctly layered volcanic rocks that have been hydrothermally altered appear on opposite side of valley (central part of photograph).

**Figure 4-8.** Boiling springs and steam vents at Bumpass Hell.

Hell is an aspect of the late stages of volcanism.

As you would suspect by comparison with Sulphur Works, rocks at Bumpass Hell have been intensely altered by hydrothermal action. In fact, the alteration processes are very much at work today. Steam and boiling water percolate unrelentingly through the ground, transforming the original rock to clays and iron oxides. The combination of hot water and steam and the presence of clay, gives rise to the numerous mud pots which bubble continually. The hydrothermal alteration processes involve large scale chemical changes. The waters are acid due to dissolved sulphur dioxide. If you should cautiously taste the water emanating from a hot spring, you will find that it tends to be sour.

As you return to the highway, note Lake Helen on the north side of the road (sign 18). Lake Helen is one of a large number of glacial lakes, as is Emerald Lake, a half mile west. These lakes have been formed by the scouring effects of glacial ice, progressively scraping out depressions which subsequently filled with water.

The vicinity offers an excellent view of Lassen Peak. Of particular interest are the prominent steep cliffs on the south side of Lassen Peak (Figure 4-9). These are composed of dacite, which is a silica-rich vari-

**Figure 4-9.** Telephoto view of south side of Lassen Peak.

ety of lava. Much of Lassen Peak is composed of masses of dacite which were pushed up out of the interior of the volcano as huge plugs of pasty, semi-molten lava which was too viscous to flow very far. Volcanoes of this type are termed volcanic domes.

**Stop 6: Lassen Peak**

A walk to the summit of Lassen Peak is a must, for the trip to the summit is both scenically and geologically rewarding. The climb involves modest effort on a well graded trail of about 2.2 miles which extends from the 8,500 foot elevation at the highway to 10,457 feet at the summit. As you start up the trail, note the banded dacite lava on your left (Figure 4-10). The bands developed as the lava flowed and subsequently cooled and congealed in an intermediate stage in Lassen Peak's development. The trail affords a succession of fine views of features on the slopes of Lassen Peak itself, including large masses of dacite (Figure 4-9) that locally project above the rock rubble that mantles much of Lassen's slopes.

Lassen Peak is a great volcanic dome that towers over the cluster of lesser domes on its south and northeast sides (Figure 4-11). The peak consists of a steep-sided core with a horizontal sectional area of about a

**Figure 4-10.** Flow banded dacite lava near foot of Lassen Peak trail.

square mile, around which is an apron of talus that extends over twice that area. Locally, on the sides of Lassen Peak, high crags of solid dacite lava project through the talus. It is probably that the talus banks that girdle Lassen came from the central dome as it was extruded upward and perhaps outward.

The relationships, as inferred by Williams, of Lassen Peak to several of the lesser dacite domes and to older rocks, are shown in the geologic section of Figure 4-12. You should keep in mind that the relationships at depth, as portrayed in this geologic section, are necessarily highly interpretative. The evidence on which the section is based has been obtained virtually entirely from exposures at the surface; there is no information from drilling, for example, that would verify the interpretations at depth. The section shows the andesite flows of the old Brokeoff volcano, which later subsided, and the dacite lavas (labeled "old Lassen Crater" in Figure 4-12) that flowed out in an intermediate stage of Lassen Peak's history and which we saw at the start of the trail. These various flows in turn, were subsequently intruded by dacite domes. The shape and size of the main dome which forms the bulk of Lassen Peak is best explained by an extrusion of pasty or semisolid lava, that was squeezed outward and upward through a pre-existing vent. The central dome lacks any definitive structure.

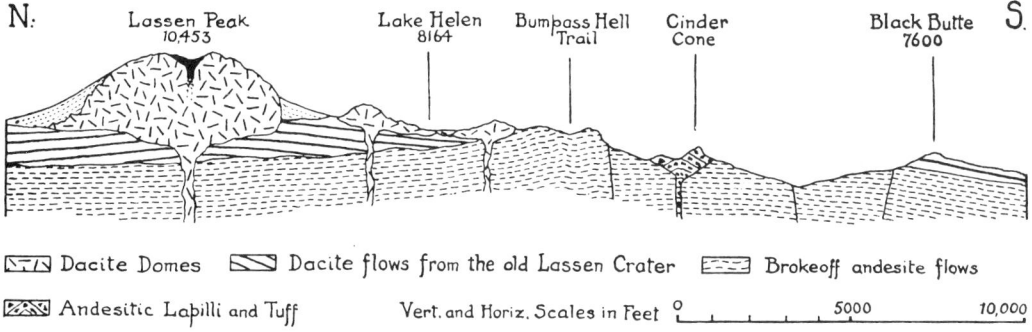

**Figure 4-11.** Aerial view southward toward Lassen Peak, showing Chaos Crags and other features of Lassen Volcanic National Park. Brokeoff Mtn. and Mt. Diller are remnants of an extensive Pleistocene volcano, now largely destroyed. Pleistocene dacite domes make up Morgan and Doe Mountains, and Eagle and Lassen Peaks. Chaos Crags are composed of recent dacite domes. Chaos Jumbles (light band, right foreground) is a volcanic avalanche that originated at Chaos Crags. (Photograph by John S. Shelton.)

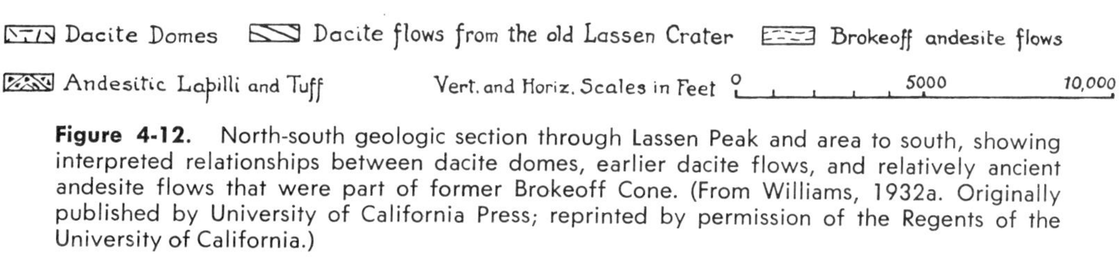

**Figure 4-12.** North-south geologic section through Lassen Peak and area to south, showing interpreted relationships between dacite domes, earlier dacite flows, and relatively ancient andesite flows that were part of former Brokeoff Cone. (From Williams, 1932a. Originally published by University of California Press; reprinted by permission of the Regents of the University of California.)

In fact, one of the most distinctive aspects of Lassen Peak is the almost complete lack of any systematic structure.

The cliffs on the south side of Lassen Peak (Figure 4-9) are smoothly polished and curve gently upward, bending upward toward the summit. Williams suggests that these surfaces may be original faces of the dome, particularly in view of the vertical striations and flutings as the faces, which rule out polishing by glacial action. While proof is difficult, it is possible that these smooth cliffs are the result of friction and polishing of the solid or nearly solid lava as it was pushed upward.

On theoretical grounds we can suppose that the core of Lassen Peak remained viscous for some time after the margins had congealed to firm rock. Furthermore, it is reasonable to speculate that the inner part contained gases and volatiles which may have imparted a vesicular character to the rock inside the dome. In addition, there may have been an ill-defined central conduit through the inner part of the dome. Williams suggests that the eruptive activities beginning in 1914 may have consisted of reheating the lining surrounding the conduit, causing it to melt and to be expelled as fluid lava, resulting in dacite flows and explosive ejections of pyroclastic material.

The rate of growth of Lassen Peak provides some interesting speculations. It is well known that volcanoes can be formed with catastrophic rapidity, as for example, Parícutin volcano in Mexico, which erupted initially in 1943 and rapidly attained its present form. Lassen Peak, however, being a volcanic dome, was formed by a different process than Parícutin. Mont Pelée on the island of Martinique, in the West Indies, may provide a better analogy. In the early part of this century, Mont Pelée acquired a central spine of glassy lava that capped the volcano. In late 1903, the spine began to rise out of Pelée's summit. Probably the spine was a mass of solidified lava that had cooled in the vent following the previous eruption of Mont Pelée. Several months after the spine began to rise, Mont Pelée erupted violently, destroying the spine, and more notably, totally destroying the city of St. Pierre, which lay at the foot of Mont Pelée, killing all but one of its 28,000 inhabitants. In those several months, the spine rose more than 100 feet. The destruction of St. Pierre resulted from an immense cloud of glowing ash and dust that settled down on the slopes of the cone and then flowed rapidly downward as a giant density current. The point to be emphasized is that extrusion of spines or domes can take place in a matter of weeks or months. Lassen Peak's dome is vastly larger than Mont Pelée's former spine, but if the rate of extrusion on Pelée is any guide, we could speculate that Lassen Peak may have attained its present height in a period of perhaps as little as five years if the extrusion and uplift was a continuous process and not interrupted by long periods of quiescence.

The main attractions of Lassen Peak are at its summit. The great panorama is breathtaking, with Mt. Shasta's 14,161 summit looming toward the northwest. Clockwise from Mt. Shasta, features within Lassen Park of direct interest to us include Chaos Crags, Devastated Area, Prospect Peak, Cinder Cone, and Butte Lake. Mountains in Nevada may be seen in the distance toward the east, and Lake Almanor and the north end of the Sierra Nevada in the middle distance toward the southeast. Completing our circular overview, we recognize Brokeoff

Mountain toward the south, beyond which the Sacramento Valley stretches to the south and west. To the west-northwest, the Trinity Alps form the prominent peaks in the Klamath region.

The view from Lassen Peak is particularly noteworthy if we consider the fact that we can see parts of six geologic provinces from this vantage point—the Cascade and Klamath provinces, the Modoc Plateau (which stretches to the north and northwest from Prospect Peak and beyond), the Great Basin in the far distance to the east, the Sierra Nevada, and the Great Valley. In fact, the only geologic province in northern California that we cannot see is the Coast Ranges province!

But now, let us turn our attention to the features on Lassen's summit. The details of features that we see in the area are, to a substantial extent, those that were formed or modified during Lassen's most recent eruptive activity, in the interval between 1914 and 1917. The first eruption occurred suddenly on May 30, 1914. For nearly a year thereafter, fragmental volcanic debris emerged intermittently from vents at the summit, where a new crater formed. Then, on May 19, 1915, molten lava entered the crater, filled it, and then poured down the western side of Lassen Peak for a thousand feet before congealing. Lava and ash also poured down the east side of Lassen Peak, where it encountered packed snow that had accumulated during the previous winter. The effect was calamitous; the mixture of snow, ash, and hot lava combined to form a mass of pasty, mud-like material that suddenly began to slip, accelerating down the east slope and devastating the forest in its path. A mile-wide swath of forest was buried under the mixture of snow, ash, and mud. Three days later, however, on May 22, the climax occurred. Folowing a brief interval of quiet, there was a violent explosion that propelled an immense cloud of volcanic ash thousands of feet into the air. The finer ash particles were drifted for long distances toward the east by the prevailing winds. Part of the blast of hot ashes and gases, however, was deflected downward over the mudflow. The effect was to reactivate the mudflow, causing further devastation, snapping additional trees like toothpicks and laying waste to an even greater area.

The blast also had a searing and erosive effect, for the cloud of hot ashes and gases flowed along the ground, scorching and abrading the remaining trees. If you look toward the northeast, you can see the path of the mudflow and hot blast, much of the path being barren of forest and appropriately named the Devastated Area.

The topography of the summit area of Lassen Peak reflects the activity of the 1914-17 era. Figure 4-13 outlines the geographic locations of the various eruptive events and shows the location of trails. By veering toward the left where the main trail branches (the right branch leads to the summit point), the 1915 flow of dacite lava can be examined. By walking still farther, the small craters that formed during 1915-16 and 1917 can be examined. The details of the dacite flow are particularly interesting (Figure 4-14). There is still some weak fumarole activity in the summit area, but it has progressively decreased during the past several decades.

*Possible Future Eruptive Activity.* Looking to the future, an obvious question is whether Lassen Peak will erupt again? No definite answer can be given, but it seems

100 • Cascade, Modoc Plateau, and Klamath Provinces

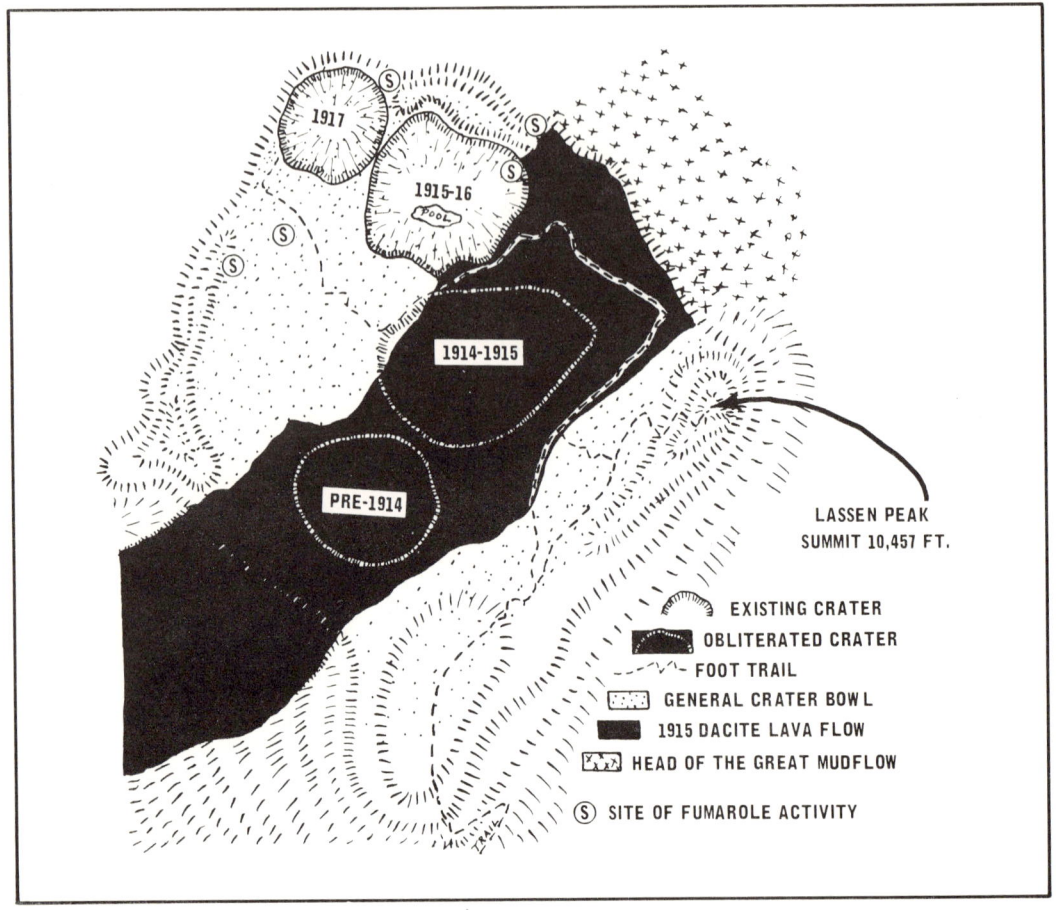

**Figure 4-13.** Map of summit area of Lassen Peak. (From U.S. National Park Service.)

reasonable to conclude that Lassen Peak is far from "dead," and that future eruptions are likely and even highly probable! The National Park Service is aware of the potential hazards from future eruptions. The U.S. Geological Survey has advised the Park Service to locate facilities away from areas of greatest potential volcanic hazards, and to monitor Lassen Peak and Chaos Crags and adjacent areas so that an impending eruption could be detected in time to evacuate people. It has been pointed out that, while some volcanoes have erupted violently without apparent warning, most eruptions are preceded by warning signs over a period of days, or even weeks or months. Geological early warning signs include (a) an increase in frequency and magnitude of local earthquakes, (b) sudden appearance of jets of steam emerging from the ground, (c) subterranean rumbling noises, (d) a marked increase in temperature and flow of hot springs and fumaroles, and (e) repeated landslides on the flanks of volcanoes. You should keep these points in mind when you visit any recently active volcanic area, including Lassen Park.

## Stop 7: Devastated Area

The stop at sign 44 provides a different view of the Devastated Area. The devastation was caused both by the mudflow, which moved down the northeast side of Lassen Peak on May 19, 1915, and by the subsequent events of May 22, 1915, in which a hot cloud of cinders, dust, and volcanic gases flowed down the mountain in the same general path, reactivating the mudflow, and doing additional damage. The mudflow of May 19 moved initially on the steep northeast slopes below the summit area, and then moved to the lowlands below (Figure 4-15). Part of the flow was divided and sent in a northward direction along Lost Creek (which the highway parallels for several miles) and the remainder was shunted in a northeasterly direction down Hat Creek. Raker Peak, which lies just to the north, served as the divider, separating the mudflow into its two branches. Figure 4-3 outlines the routes of the mudflow.

**Figure 4-14.** Summit area of Lassen Peak. Dacite flow of 1915 is in foreground and crater formed by subsequent eruptive activity is toward the rear. Note people near middle of view for scale.

**Figure 4-15.** Photograph of northeast side of Lassen Peak taken in 1920 showing part of area devastated by flows on May 19 and May 22, 1915. Flows spread over a broad area to left of Crescent Crater (in right middle foreground). Since photograph was taken, natural reforestation has obscured many dead trees.

The mudflow left a wide swath through the forest, pushing over trees so that their fallen trunks point in the direction of movement. Fertile meadows were covered with up to 20 feet of mud and debris. The hot blast of May 22 completed the devastation by breaking off and pushing over additional trees, and scorching and abrading those tree trunks that remained standing. Since 1915, the Devastated Area has undergone slow natural restoration. The young trees that replace those destroyed were not planted by man, but represent natural reforestation. Meanwhile, the trunks of fallen trees are gradually decaying.

**Stop 8: Raker Peak**

Raker Peak (sign 45) is a volcano that rises about 1,200 feet above its immediate surroundings. Raker Peak is a composite volcanic feature. Most of Raker Peak is a shield volcano composed principally of andesite. Late in its history, however, a stiff dome of dacite lava extended along its south end. It is thus similar to Lassen Peak in that it has a dacite dome, but it differs in that the rest of Raker Peak is more typically a shield volcano.

**Stop 9: Hot Rock**

Hot Rock (sign 48) is one of many large boulders carried down from Lassen Peak in the mudflow of May 19, 1915. Hot Rock is a mass of dacite weighing about three hundred tons which was derived from the dacite flow that immediately preceded the mudflow on May 19, 1915. When the dacite flow occurred on May 19, the dacite lava was molten. Although the lava congealed quickly, large blocks such as Hot Rock, remained sufficiently hot after being transported by the mudflow that they were capable of starting fires. Weeks elapsed before Hot Rock finally cooled, thus the name "Hot Rock." Close examination reveals that the dacite includes many inclusions of pumice.

**Stop 10: Chaos Crags**

Chaos Crags (Figure 4-16) lie about 1½ miles southeast of the view point at sign 63. The Crags are formed from four or more dacite plug volcanoes. These dacite plugs are similar to the plug dome that forms the bulk of Lassen Peak, except those of Chaos Crags are much younger, being less than 1,200 year old. Although smaller, the Chaos Crags dacite plugs were probably formed like that of Lassen Peak, being squeezed upward as pasty masses, perhaps like toothpaste squeezed from a giant tube.

Our attention is drawn to the chaotic jumbles of angular blocks, appropriately named Chaos Jumbles, which extend over the area between the highway and Chaos Crags. The manner in which the blocks were transported has evoked considerable interest. There is little doubt that the blocks came from Chaos Crags, for they consist of dacite that originated in the northern plug dome of Chaos Crags, and were set free as a result of a series of explosions that occurred near the base of the plug dome. The blocks were transported for distances of up to a mile and a half by a series of avalanches. The avalanches appear to have moved with very high velocities, for the loose masses of blocks had sufficient momentum to move uphill on the south flank of Table Mountain, rising as much as 400 feet elevation. Figure 4-11 reveals the flow lines in the succession of blocks forming Chaos Jumbles.

The remarkable mobility of these avalanches raises the question of how they were transported. It is unlikely that they moved as mudflows, although transport by mudflow has certainly been an important transport process elsewhere in the region, as in the Devastated Area for example. Instead, it seems more likely that the masses of broken rock moved as massive avalanches in which momentarily trapped compressed air served as a cushion beneath the rock mass, greatly reducing friction and permitting it to glide over the land surface at speeds of more than 100 miles per hour.

## MODOC PLATEAU PROVINCE

The transition to the Modoc Plateau Province within and adjacent to Lassen Park is not as obvious as you might initially suspect. Nevertheless, the differences will be noticeable as you cross into the Modoc country. Our route is via Highway 89, through the entrance in the northwest corner of the park. Highway 89 and 44 join just outside the park and coincide for about 14 miles toward the northeast. As you drive northeast toward Old Station, note the modest topographic scarp at Big Spring (in the lower left corner of Figure 4-17). This scarp is a fault scarp and is more or less representative of the many fault scarps that are among the distinguishing features of the Modoc Plateau province. The boundary be-

**Figure 4-16.** Chaos Crags. Chaos Jumbles in foreground consist of angular dacite blocks transported by avalanches from Chaos Crags. Lassen Peak appears in distance on right.

tween the Cascade and Modoc provinces is necessarily somewhat arbitrary, but we could regard this point as a transition. However, a much more prominent fault scarp lies ahead at Hat Creek Rim.

**Stop 11: Subway Cave and Hat Creek Rim**

Subway Cave is one of a large number of lava caves in Hat Creek Valley (Figure 4-17). These lava caves occur in relatively young basaltic lavas (they are probably not over 2,000 years old) which flowed over the pre-existing valley. The caves are an effect of uneven cooling as the lava flow spread over the valley floor. The surface of the lava flow congealed, but the interior remained molten due to its higher temperature. As a result, the fluid interior lava continued to flow through a network of large tubes, leaving behind walls and ceilings of congealing lava.

The entrance to Subway Cave is located at one of the places where the roof over the lava tube has collapsed. The interior of Subway Cave reveals the configuration of the last flow through Subway Cave. The wrinkled flow lines in the floor indicate that the lava was in a thick, pasty condition just before it became too viscous to move further. The walls and ceiling of the cave contain small glassy "stalactites" that may have

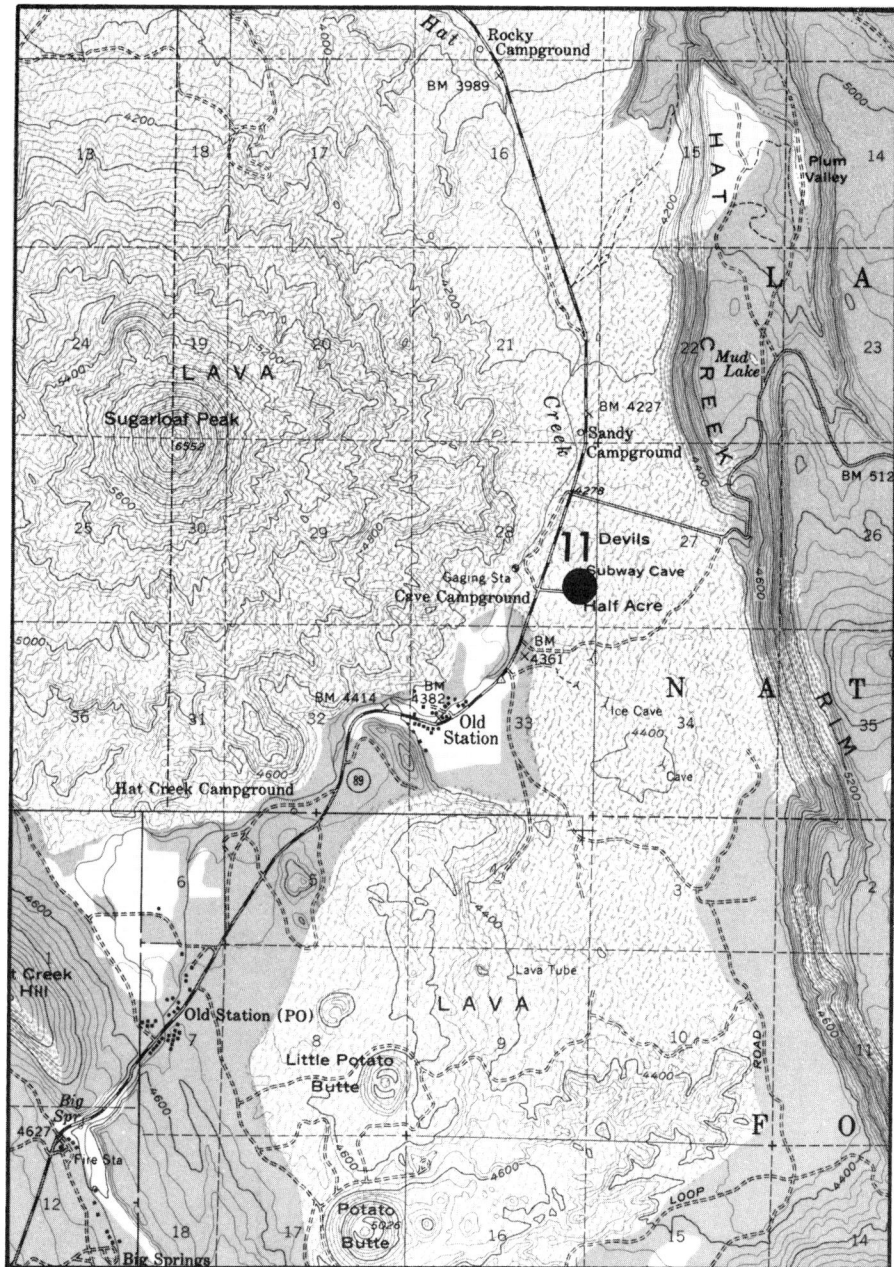

**Figure 4-17.** Hat Creek valley, containing Subway Cave. Hat Creek Rim is a prominent fault scarp that bounds the valley on east. Sugarloaf Peak is a shield volcano. East-west width of map is 5 miles. (From U.S.G.S. Prospect Peak 15' topographic quadrangle.)

formed when hot gasses remelted the walls of the cave.

The lava flow that contains the caves and forms the present floor of Hat Creek Valley poured out from a series of north-south trending fissures located about 1½ miles south of Old Station, on a low ridge which can be identified on the topographic map (Figure 4-17) by looking immediately east and northeast of the word LAVA. Scattered over this low ridge is a belt of spatter or driblet cones that are roughly aligned in a north-south direction, and which range in height from 6 to 30 feet. These cones probably represent the accumulation of jets of lava hurled from a series of lava fountains located along the fissures, from which the Hat Creek lava flow emerged. Similar spatter cones are prominent in Lava Beds National Monument, about 75 miles to the north.

Highway 44, which is our route to Cinder Cone, extends eastward from Hat Creek Valley, rising about 800 feet as it climbs Hat Creek Rim. Hat Creek Rim is a fault scarp, or more correctly a scarp resulting from a complex of interconnected faults which are subparallel or *en echelon* with respect to each other. The present topography (Figure 4-17) partly reflects the displacement and relationships between the fault blocks. Individual fault blocks form inclined ramps, the highway route taking advantage of the ramps to ascend the otherwise formidably steep rim (Figure 4-18).

**Stop 12. Cinder Cone**

The northeast corner of Lassen Park provides some scenic and geologic surprises. In fact, many people consider that a visit to Cinder Cone (Figure 4-19) is the most rewarding experience of all in Lassen Park. Cinder Cone is a very young cone of basaltic cinders which lies adjacent to equally young basaltic lava flows, known as the Fantastic Lava Beds. Park in the campground at Butte Lake and walk along the old Emigrant Trail to Cinder Cone. Pick up a self-guiding trail leaflet as you leave the campground. The trail skirts the edge of

**Figure 4-18.** Hat Creek Rim looking north-northwest. Irregular surface in the foreground is surface of lava flow extending over floor of Hat Creek Valley.

Fantastic Lava Beds and provides an excellent view of the edge of a basaltic flow that gradually ceased to move as it congealed. The basalt flow probably moved as a pasty, slow-moving mass that was red-hot inside, but relatively cool on the outside. As the flow advanced, the congealed lava crumbled off, forming piles of the irregular blocks of lava that mantle the flow.

The basalt is unusual in that it contains glassy grains of quartz. Ordinarily, basalt does not contain free quartz because its gross chemical composition contains insufficient silica to yield quartz upon crystallization. It is possible, however, that the quartz may have been derived by mixing of magmas, in which silica-rich dacitic magma was incorporated with basaltic magma.

The trail branches near the foot of Cinder Cone, the left trail leading to the summit. The slopes of Cinder Cone are mantled with fine cinders, providing the impression that much of the cone is composed of cinders. The slope of the cone is about 35 degrees, which is the angle of repose of loose fine cinders, and the cone is almost perfectly symmetrical, suggesting that the cinders and ash erupted from a central vent. The bulk of the ejected material fell uniformly around the vent, suggesting the lack of a strongly prevailing wind direction and accounting for the symmetry of the resulting cone. The lack of erosion of Cinder Cone is explained both by its extreme youthfulness and the fact that the porous cinders absorb even the heaviest rainfall, preventing the water from running off the surface and eroding it in ordinary fashion.

The summit of Cinder Cone reveals a pronounced depression, or crater, and provides a spectacular panoramic view. Prospeck Peak, immediately to the northeast, is an excellent example of a symmetrical shield or Hawaiian-type volcano, being composed of basalt which flowed quickly from a central vent. The small cinder cone on top of Prospect Peak indicates that its final activity was more explosive.

The view toward the east from the rim of Cinder Cone reveals the spectacular lava flows that formed late in Cinder Cone's history. There are several generations of flows. The flows that appear lighter in Figure 4-19 are covered with volcanic cinders which have been oxidized by steam and hot water to reddish and buff colors. They are called the Painted Dunes. The flows that appear darker are younger and probably formed in the eruption of 1851. The flow lines are readily visible.

Descend from Cinder Cone, by taking the trail that extends in a southerly direction from the east side of the rim. If you turn off the trail for a short distance when you are about three-quarters of the way down from the rim, you can actually look inside a vent from which lava poured, probably in 1851. You can trace the path of the lava as it flowed from this vent toward Snag Lake and subsequently veered off to the left toward Butte Lake.

The trail curves around the south edge of Cinder Cone, rejoining the Emigrant Trail. Near the trail junction, there is a spectacular display of volcanic bombs of various sizes (Figure 4-20) which were ejected from Cinder Cone.

Because of Cinder Cone's extreme youthfulness, its history is known relatively well. Its initial development, including the ejection of cinders to build the cone, as well as outpouring of fluid lavas from near its base, probably began less than 2,000 years ago. This estimate, an educated guess at

**Figure 4-19.** View east toward Cinder Cone. The earliest flows (light-colored area) are quartz basalts, probably less than 2,000 years old. Youngest flow (dark tongue in center of light-colored area) also consists of quartz basalt and probably was extruded in 1851. Double rim of Cinder Cone indicates that more than one interval of explosive activity occurred during its development. Photo by John S. Shelton.

best, is based on the freshness of the lava flows and the lack of erosional modification of the cone. Cinder Cone's double rim suggests that the development of the cone involved at least two periods of eruption. An early lava flow from the base was subsequently mantled by cinders of a later eruption. One of the early flows dammed the creek that flows into Butte Lake, forming Snag Lake.

It is probable that the last eruption occurred in 1851. During the winter of 1850-51, residents of the Sacramento Valley observed the distant light of a great "fire" east of Lassen Peak. Because of the presence of hostile Indians, however, no attempt was made to investigate the cause of the light. Two prospectors who traversed through the general vicinity of Cinder Cone in the summer of 1851 reported passing an active volcano, thus strengthening the assumption that the great fires were the eruptions of Cinder Cone.

### Stop 13: Burney Falls

McArthur-Burney Falls Memorial State Park is one of the oldest units of the California State Park system. The falls themselves are one of the system's crown jewels. Burney Falls are formed where Burney Creek plunges over a 129-foot escarpment as a twin fall, accompanied by numerous springs that issue plumes of water from the walls of the cliff (Figure 4-21). Burney Falls lies in the border zone between the Cascade Range and the Modoc Plateau provinces. We shall regard it as lying within the Modoc Plateau for convenience' sake.

The steady flow of Burney Creek at Burney Falls, even in the summer after months of little or no rain, is explained by the fact that Burney Creek is largely fed by springs. In fact, much of the water in Burney Creek downstream from the falls is spring water that has emerged at the falls. The ultimate source of these springs is surface water which has fallen as rain, or was derived from melted snow, and has subsequently percolated into the ground in the area that is tributary to Burney Creek (Figure 4-22). Much of the basalt is highly permeable to the flow of water, so that instead of moving in Burney Creek and its branches,

**Figure 4-20.** Volcanic bombs ejected from Cinder Cone.

# 110 • Cascade, Modoc Plateau, and Klamath Provinces

**Figure 4-21.** Burney Falls.

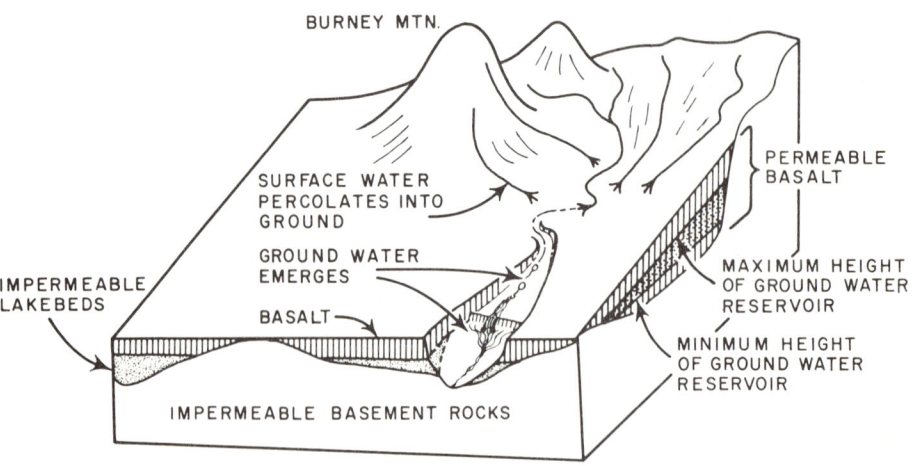

**Figure 4-22.** Diagram showing relations of springs to underground geology at Burney Falls and area to south. Burney Mountain is about 15 miles south of Burney Falls. (From Aune, 1964.)

much of the water moves underground until it emerges at springs.

The basalt which conducts the water at Burney Falls rests unconformably on some older lake bed deposits, which in turn rest upon older volcanic pyroclastic rocks (labeled basement rocks in Figure 4-22). Both the lake bed deposits and the basement rocks are relatively impermeable, whereas the overlying basalt is highly permeable. The effect is to cause part of the water which moves through the basalt to be confined immediately above the lake beds. When the contact of the lake beds and the overlying basalt is exposed in a cliff, as at Burney Falls, the water emerges as a line of springs along the contact.

## CASCADE PROVINCE

### Stop 14: Burney Mountain

Stop in the parking area adjacent to the highway vista point (designated by signs). Here you can see Burney Mountain toward the southeast. Burney Mountain is a volcano that rises about 4,000 feet above its base and is roughly conical in shape. The rocks exposed on Burney Mountain are andesite flows, but it is possible that Burney Mountain has an inner core of pyroclastic material similar to that of Magee Mountain, farther south. Burney Mountain appears to be very little modified by erosion and has not been glaciated. It is relatively young, perhaps less than 10,000 years old.

Magee Mountain, which is at a greater distance and appears to the right of Burney Mountain, is another volcano that also rises about 4,000 feet above its base. Unlike Burney Mountain, however, Magee Mountain has been sculptured by glaciation and is judged to be considerably older, probably Pleistocene. Magee Mountain consists of andesite flows that overlie a core of volcanic breccia. Both Burney Mountain and Magee Mountain are part of the loose chain of peaks which forms the High Cascades. An indefinite boundary between the Cascade and Modoc Plateau provinces lies east of the two mountains.

### Stop 15: Montgomery Creek Formation

This stop involves inspection of the Montgomery Creek Formation in extensive road cuts along Highway 299. Stop where the road direction changes from northwest-southeast, to east-west, roughly 1½ highway miles east of the village of Montgomery Creek. The Montgomery Creek Formation consists of lensing conglomerates, cross-bedded sandstones, and coal beds. The conglomerates contain rounded cobbles up to 6 inches or more in diameter which are composed of a variety of types of volcanic rock.

The Montgomery Creek Formation was deposited by streams during the Eocene Epoch. The formation in turn rests unconformably on sedimentary rocks deposited in shallow seas in the latter part of the Cretaceous Period. Locally, the Montgomery Creek Formation is overlain unconformably by Pliocene volcanic rocks. The Montgomery Creek Formation can be related to both the rocks of the Great Valley province and to the Cascade province. The Cretaceous strata on which it rests are widespread in the Sacramento Valley (the northern part of the Great Valley province), whereas the volcanics of the Cascade Province stratigraphically succeed the Montgomery Creek Formation.

We can envision the conditions under which the Montgomery Creek Formation was deposited. Streams draining highland areas in which volcanoes were active, contributed material to the deposits. The streams flowed over low plains, flooding intermittently. Swamps in which plant material was preserved formed locally, creating thin seams of low-grade coal or lignite.

## KLAMATH PROVINCE

The Klamath province extends over about 12,000 square miles in northwestern California and southwestern Oregon. This province is a region of rugged topography and is a remote and sparsely populated terrain.

The rocks of the Klamath province can be divided into two broad categories, namely stratified sedimentary and volcanic rocks, and plutonic intrusive rocks. The stratified rocks typically consist of a variety of sedimentary rocks, including graywacke siltstones and sandstones, limestones, tuffs, volcanic breccias, and lava flows. The siltstones and sandstones, however, being of graywacke type, are quite different from ordinary siltstones and sandstones. We usually think of sandstone consisting of sorted, subrounded grains of quartz and perhaps feldspar. Graywacke sandstone, by contrast, tends to be ill-sorted and to be composed of angular grains of a variety of materials, including fragments of volcanic rocks and of lithified clay. Graywacke sediments are "youthful" or "immature" as sediments are concerned. The close association of graywackes with volcanic pyroclastic material, and the general occurrence of graywackes in regions of complexly deformed structure, suggests that their origin is intimately connected with the evolution of the geological provinces in which they occur.

The stratified rocks of the Klamath province range in age from Ordovician to Tertiary. It is possible that stratified rocks older than Ordovician are present, but they have not been recognized. Rock formations of the Klamath province can be placed logically in three principal belts, the eastern, central, and western belts, whose extent is shown in the geological map of Figure 4-23.

The eastern Klamath belt contains a sequence of rock formations that range in age from Ordovician to Jurassic and dip generally toward the east, although the strata are locally folded. Eastern belt formations have an aggregate thickness of perhaps 40,000 to 50,000 feet, and are relatively unmetamorphosed, except adjacent to some igneous intrusions, where thermal metamorphism has occurred.

Rocks of the central belt have been transformed to schists, which represent former sedimentary and volcanic rocks that have been altered by regional dynamic metamorphism. The rocks of the central metamorphic belt are separated from those of the eastern belt by masses of ultrabasic (also termed ultramafic) rocks that consist chiefly of peridotite which has been converted to serpentine. The peridotite occupies belts that roughly parallel the general grain of the region. On the west, however, the rocks of the central metamorphic belt are separated from the western Klamath belt by large faults.

The rocks of the western belt are complexly deformed and consist of graywacke, siltstone, chert, dark volcanic rocks, and lenses of coarsely crystalline limestone. These rocks also include ultrabasic rocks and have been intruded on a large scale by

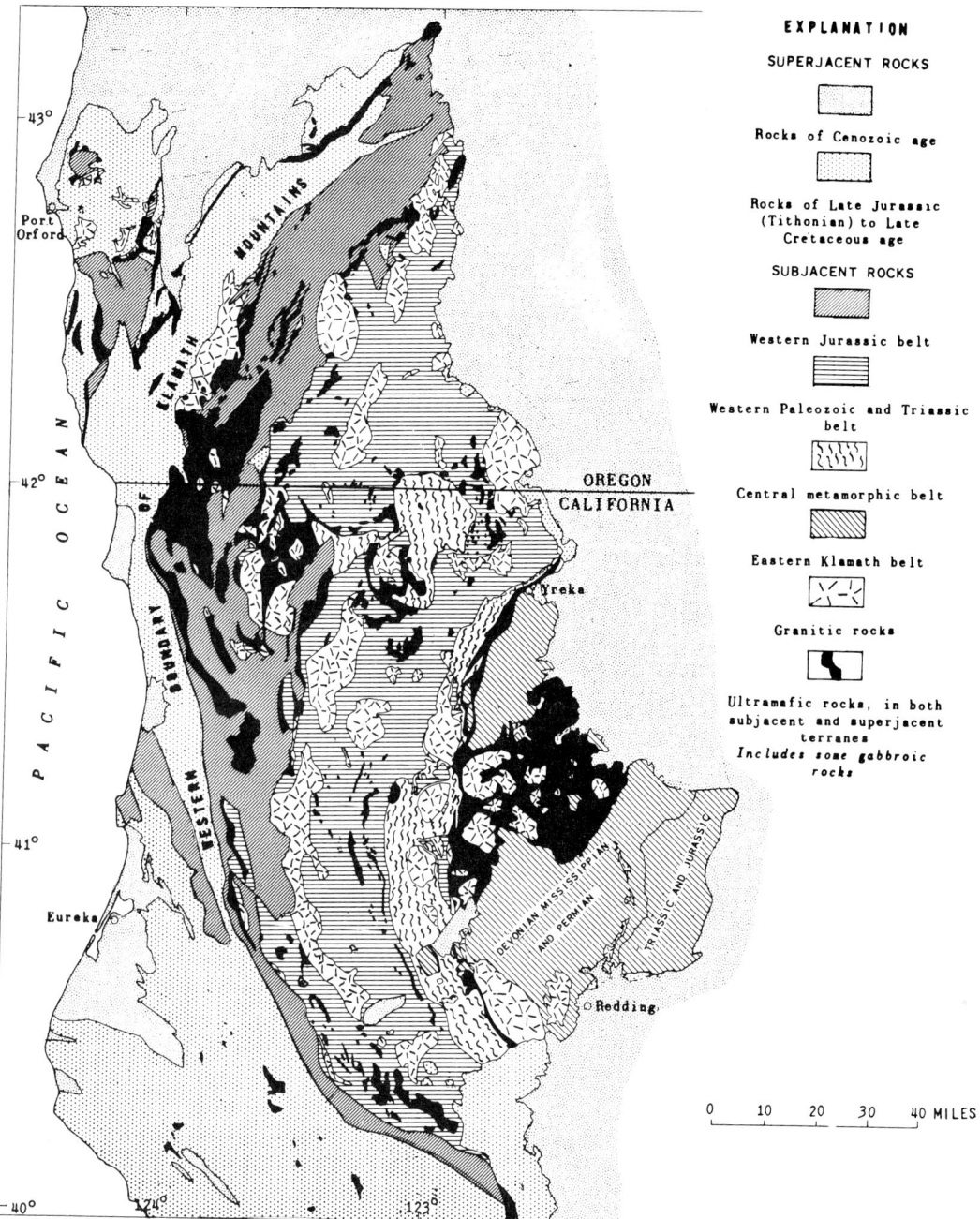

**Figure 4-23.** Generalized geologic map of northwestern California and southwestern Oregon showing main belts of Klamath province. (From Irwin, 1966.)

granitic rocks. Fossils that have been found provide a basis for separating the western belt into a Paleozoic and Triassic subbelt, and a Jurassic subbelt.

You will pass outcrops of a variety of rocks of the eastern Klamath province as you travel on California 299 toward Redding, and along Interstate 5, from Redding northward toward Castle Crags.

**Stop 16: Castle Crags**

Castle Crags State Park is another jewel of the California State Park system (Figure 4-24). While small in area, Castle Crags Park is large on scenery, including a splendid view of Mt. Shasta. The crags themselves are immense weathered blocks of granitic rock. Many of the fractures or joints which define the blocks are nearly vertical and have guided the weathering of the blocks to form the sharp crags.

Take the park trail into the heart of the crags. Notice that the granodiorite in the crags contains large individual crystals of light gray feldspar up to an inch long, in many places. Biotite and some quartz are also present, but in smaller proportions. You will probably notice veins of white quartz up to six inches wide that cut across the granodiorite. On the trail before you reach the crags, you may notice loose fragments of serpentine. The granodiorite intrusion, from which the crags are formed, is surrounded by ultrabasic rocks, consisting mostly of serpentinized peridotite. The geologic maps of Figures 4-2 and 4-23 show the general relationships.

The age of the granodiorite is probably about 170 million years. However, the ages of the ultrabasic rocks are less well known. The problem of determining the ages of the ultrabasic rocks is compounded by the fact

**Figure 4-24.** Castle Crags.

that serpentine is apparently capable of flowage while cold. Thus, it seems possible that the serpentine masses have continued to flow for long intervals of time, and may still be doing so. On the other hand, the ultrabasic rocks (before or after serpentization) may have been emplaced by large scale fault movements. If you wish an additional look at the ultrabasic complex, drive westward for several miles along the Forest Service road that joins Interstate 5 at Castella Interchange (which is the interchange that serves Castle Crags Park).

## CASCADE PROVINCE

### Stop 17: Mt. Shasta

Most connoisseurs of California scenery would probably nominate Mt. Shasta as California's most outstanding mountain peak (cover photograph). Although Mt. Whitney is several hundred feet higher than Shasta's 14,162 feet, no mountain in California exceeds Shasta for sheer relief and splendid isolation. Mt. Shasta is, of course, part of the chain of peaks that form the High Cascades. In fact, Mt. Shasta is the largest of all of the Cascade volcanoes. From a base of about 17 miles in diameter, it rises more than 10,000 feet above its surroundings, encompassing a volume of about 80 cubic miles of volcanic rock.

A trip to the south flank of Mt. Shasta is a memorable experience. Take the paved road to Panther Meadows. The road begins in the town of Mt. Shasta and ends at the ski area. A good vantage point for viewing Mt. Shasta is at the last sharp turn before reaching the ski area.

The shape of Mt. Shasta is that of two cones that partly coalesce, Mt. Shasta itself and its satellite cone Shastina (Figure 4-25). The slopes of Mt. Shasta and Shastina vary from a maximum of about 35° in the summit areas, to about 5° or less near the base. The main cone of Mt. Shasta is so young that only its outermost part is exposed by erosion. The exposed rocks consist largely of massive andesitic flows that average about 50 feet thick and appear to have emanated from a single vent. Some pyroclastic debris is present, but in smaller proportion. The slopes of Mt. Shasta are generally governed by the dips of the lava flows themselves. In places where the glacial erosion has cut into the volcano, dipping flows can be observed.

Mt. Shasta is a youthful volcano with a potential for renewed eruption, judging from the evidence of its activity in the recent past. Late in Shasta's history, a fissure opened across Shasta's cone in a north-south direction. From this fissure, a series of eruptions formed a chain of small domes and cinder cones. At the base of Mt. Shasta along its south side, Everitt Hill consists of a small shield volcano with a small cinder cone at its summit. Flows of andesitic lava from this volcanic vent flowed down the canyon of the Sacramento River for more than 40 miles, where they may be seen along Interstate 5 north of the Shasta Lake area. They consist of columnar-jointed lava which, in places, rests upon stream gravels.

Also late in Shasta's history, possibly at about the same time as the development of the north-south fissure, an east-west fissure opened on the west flank of the volcano. Eruptions along this fissure built a small lava-and-cinder cone a mile west of the summit, and then soon afterward, other flows poured from a vent a half mile farther west, progressively building the cone of

**Figure 4-25.** Mt. Shasta and Shastina viewed from the north-northwest.

Shastina. The last eruptions of Shastina built several small domes within its crater which formed after the extensive glaciation that occurred in late Pleistocene time. The evidence consists of the fact that these domes are unmodified by ice action, whereas most of Mt. Shasta and Shastina have been extensively glaciated.

Glacial processes are still active on Mt. Shasta inasmuch as there are small glaciers around its summit. The aggregate volume of present-day glaciers, however, is much less than the glaciers which formerly covered Mt. Shasta. Furthermore, there is evidence that the glaciers have retreated greatly in the past century.

As you drive along northward from the town of Mt. Shasta, you will notice Black Butte, an isolated volcanic cone on the west-southwest edge of Mt. Shasta. Interstate 5 passes close to Black Butte, as does the road to Panther Meadows. Black Butte is almost perfectly conical in shape, being about 2,500 feet high and 1½ miles in diameter at its base. Black Butte is a volcanic dome consisting of andesite. It was apparently extruded as a pasty mass of lava, from which blocks of lava crumbled off to yield masses of breccia that completely mantle the solid core of the dome, except for a few crags at the summit.

The latest eruptions of Mt. Shasta appear to have occurred at the summit vent of the main cone. They produced layers of pumice and cinders containing blocks and bombs of glassy andesite. In places, the rock has been altered by hot spring activity to a deep reddish brown color, forming the Red Banks (a locality familiar to climbers who ascend from Mt. Shasta's south side) on the south side of the summit crater. The latest explosion which covered the upper part of the mountain with a thin layer of brown pumice, may have occurred in 1786, when clouds of ash from a distant eruption were observed by mariners cruising offshore along the coast of northern California.

## BIBLIOGRAPHY

ANDERSON, C. A. 1940. Hat Creek lava flow. *Am. Jour. Sci.* vol. 238, no. 7, pp. 477-92.
  Provides an excellent account of the Hat Creek lava flow (Stop 11).

AUNE, Q. A. 1964. *A trip to Burney Falls*. Mineral Information Service, California Div. of Mines and Geol., vol. 17, no. 10, pp. 183-191.
  A well written description and interpretation of the geology surrounding Burney Falls.

DAVIS, G. A. 1966. Metamorphic and granitic history of Klamath Mountains, in *Geology of Northern California*, E. H. Bailey, ed. California Div. of Mines and Geol. Bull. 190, pp. 39-50.
  Reviews problems in regional studies of metamorphic and plutonic rocks of Klamath province.

EVANS, J. R. 1963. *Geology of some lava tubes, Shasta County*. Mineral Information Service, California Div. of Mines and Geol., vol. 16, no. 3.
  Describes Subway Cave and other lava caves in Hat Creek Valley. 7 pp.

FINCH, R. H., and ANDERSON, C. A. 1930. The quartz basalt eruptions of Cinder Cone, Lassen Volcanic National Park, California. Univ. of California Pubs., Bull. of Dept. of Geol. Sci. 19: 245-73.
  Discusses eruptive history of Cinder Cone, as well as providing a detailed analysis of problems of the presence of crystals of quartz in otherwise low-silica basalt.

IRWIN, W. P. 1966. Geology of the Klamath Mountains province, in *Geology of Northern California*, E. H. Bailey, ed. California Div. of Mines and Geol. Bull. 190, pp. 19-38.
  Authoritative overview of Klamath province by a geologist who has worked for many years in the Klamath region.

LOOMIS, B. F. 1966. *Pictorial history of the Lassen volcano*. Loomis Museum Association, Lassen Volcanic National Park.
  Excellent account of Lassen's eruptive activity in 1914-17 era. Contains classic photographs and detailed chronology of events by pioneer volcanic observer Loomis (can be purchased in Lassen Park). 100 pp.

MACDONALD, G. A. 1963. Geology of the Manzanita Lake quadrangle, California. U.S. Geol. Survey Map GQ-248.
  Colored geologic map of area embracing northwestern part of Lassen Park and extensive area outside the park. Scale 1:62,500.

———. 1964. *Geology of the Prospect Peak quadrangle, California*. U.S. Geol. Survey Map GQ-345.
  Colored geologic map of area embracing northwestern part of Lassen Park and area to the north, including part of Hat Creek Valley. Scale 1:62,500.

———. 1966. Geology of the Cascade Range and Modoc Plateau, in *Geology of Northern California*, E. H. Bailey, ed. California Div. of Mines and Geology, Bull. 190, pp. 65-96.
  Summarizes geology of these two provinces, with extensive discussion of Lassen Peak.

MACDONALD, G. A. and KATSURA, T. 1965. Eruption of Lassen Peak, Cascade Range, California, in 1915: Example of mixed magmas: Geol. Soc. Am. Bull. 76: 475-82.
  Describes unusual lithology of pumice erupted from the Lassen Peak in 1915. Pumice is banded; dark bands composed of lava of andesitic composition, and light bands of dacitic lava. These two quite different types of lava appear to have been derived from two magmas. The manner in which they were mixed together is highly speculative.

SCHULZ, P. E. 1959. *Geology of Lassen's landscape*. Loomis Museum Assn., Lassen Volcanic National Park.

Readable, general introduction to geology from point of view of a visitor to Lassen Park (can be purchased in Lassen Park). 98 pp.

———. 1968. *Road guide to Lassen Volcanic National Park.* Loomis Museum Assn., Lassen Volcanic National Park.

A colorful, well written general guide to sights along Highway 89 within Lassen Park. It is keyed to the numbered roadside signs (can be purchased in Lassen Park). 40 pp.

WILLIAMS, H. 1932 A. *Geology of the Lassen Volcanic National Park, California.* Univ. of California Pubs., Bull. of Dept. of Geol. Sci. 21: 195-385.

Classic, detailed account of geology of Lassen Park.

———. 1932 B. Mount Shasta, a Cascade volcano: *Jour. Geol.* 40: 417-29.

Well-written description of Mt. Shasta.

———. 1934. Mount Shasta, California. *Zeitschrift fur Vulkanologie.* Band 15, pp. 225-53.

Second of two definitive papers on Mt. Shasta by Howel Williams. Contains numerous photographs of Mt. Shasta.

———. 1969. Geology of Lassen Volcanic National Park, California, in *Geologic Guide to the Lassen Peak, Burney Falls, and Lake Shasta area, California,* R. A. Matthews and J. L. Burnett, eds. Annual field trip guidebook of the Geological Society of Sacramento-1969.

An abridgement of Williams 1932 article. Same guidebook also contains description of Lassen Peak's eruptive activities in 1914-17 era (available through the University of California at Davis, Department of Geology, price $5.00).

*Appendix*

# The Geological Time Scale

| Era | Period | Epoch | Tentative Absolute Age |
|---|---|---|---|
| Cenozoic | Quaternary | Holocene | 11,000 years |
| | | Pleistocene | 2 million years |
| | Tertiary | Pliocene | 12 |
| | | Miocene | 26 |
| | | Oligocene | 37 |
| | | Eocene | 53 |
| | | Paleocene | 70 million years |
| Mesozoic | Cretaceous | | 135 |
| | Jurassic | | 190 |
| | Triassic | | 230 million years |
| Paleozoic | Permian | | 280 |
| | Pennsylvanian | | |
| | Mississippian | | 350 |
| | Devonian | | 400 |
| | Silurian | | 430 |
| | Ordovician | | 500 |
| | Cambrian | | 600 million years |
| Precambrian | | | 600-3600 million years |
| ----------Lost Interval---------- | | | |
| Origin of Earth | | | 4600 million years |

# Indexes

## INDEX OF GEOLOGIC FEATURES

Andesite, 85, 92, 115
Ash, volcanic, 82, 99, 101, 106, 116
Avalanches, 97, 103

Basalt, 104, 107, 109
Basement complexes, 2, 11, 16, 18, 25, 44, 74, 112
Batholiths, 44, 51, 55, 74, 78, 113
Bay-mouth bar, 7
Beaches, 7, 13
Bombs, volcanic, 107, 109
Boring clams, 41
Burrowing organisms, 19, 30, 32

Calderas, 91
Caves, 104
Chert, 5, 6, 11
Clam borings, 30, 32, 41
Climate, 43, 80
Cones, volcanic, 81, 91, 106, 115
Conglomerates, 16, 19, 23, 29, 32, 36, 111
Continental shelf, 2, 5
Coulees, 82
Craters, volcanic, 81, 91, 100, 106

Cross stratification, 26, 29, 35, 76, 111

Dacite, 85, 94, 96, 102
Diatomaceous siltstone, 38
Dikes, 26, 39, 58, 70
Diorite, 56, 60, 71
Displacement of faults, 38, 44, 106
Domes, exfoliation, 60, 69, 71, 72
Domes, volcanic, 95, 103, 116
Dumped deposits, 38
Dune sand, 13, 34

Earthquakes, 7, 10, 100
Erosion, 5, 7, 19, 30, 34, 41, 54, 80, 92, 94, 111
Eruptions, volcanic, 5, 82, 98, 104, 107, 115
Estuaries, 7, 12, 35
Evolution, topographic, 10, 62, 82, 104
Exfoliation, 60, 69, 71, 72

Faults, 2, 6, 10, 14, 27, 33, 38, 43, 91, 103, 112
Fault scarps, 43, 103
Fault zone topography, 7, 10, 33, 105

Field trip route maps, xiv, 8, 10, 12, 15, 20, 22, 31, 35, 40, 46, 47, 52, 86, 89, 100, 105
Folds, 6, 11, 24, 29, 32, 49, 76
Fossils, 14, 30
Flow of glaciers, 67
Fumaroles, 88, 93, 99

Geologic maps, xi, xiii, 9, 21, 46, 54, 57, 87, 113
Geologic provinces, xi, 1, 43, 80, 84, 103, 111, 112, 115
Geologic time scale, 2, 119
Glacial erosion, 65, 68, 92, 94, 111, 115
Glacial moraines, 58, 66, 69, 72, 77
Glacial outwash, 48
Glacial polish, 66, 68, 72, 75
Glacial stages, 66
Glacial stairways, 66, 70
Glacial tills, 66, 77
Glaciation, 65
Glass, volcanic, 82
Granitic-metamorphic core complex, 3
Granitic rocks, 3, 16, 26, 43, 50, 55, 58, 78, 113
Granodiorite, 25, 29, 55, 114
Gravels, 48, 74

121

Graywackes, 4, 19, 24, 35, 76, 112
Greenstones, 5, 45

Hanging valleys, 59, 65
Hot springs, 88, 92, 94, 116
Hydrothermal alteration, 88, 91, 94, 116

Ice, glacial, 67
Intrusive rocks, 3, 11, 16, 25, 39, 50, 55, 58, 70, 76, 78, 95, 112

Joints, 54, 60, 68, 70, 76, 114

Lakes, 59, 64, 69, 81, 94
Landslides, 23, 38, 101
Lateral moraines, 77, 80
Lava caves, 104
Lavas, 5, 45, 82, 85, 88, 92, 96, 104, 115
Limestones, 81

Maps, geologic, xi, xiii, 9, 21, 46, 52, 57, 87, 114
Maps, route, xiv, 8, 10, 12, 14, 15, 20, 22, 31, 35, 40, 86, 89, 90, 105
Maps, topographic, xi, 10, 12, 14, 15, 22, 31, 35, 40, 52, 90, 106
Marine terraces, 13, 23, 25, 28, 31, 34, 38, 41
Marshes, 12, 35
Metamorphic rocks, 3, 44, 74, 112
Moraines, 58, 66, 69, 72, 77, 80
Mudflats, tidal, 7
Mudflows, 99

Mudstones, 5, 19

Natural bridges, 40

Obsidian, 82
Ocean-floor spreading, 5

Pegmatite dikes, 58, 70
Peridotite, 112, 114
Plates, tectonic, 5
Pillow lavas, 5
Plucking, glacial, 68, 70
Plutons, 3, 25, 50, 55, 78
Polish, glacial, 66, 68, 72, 75
Potassium-argon age dates, 4, 25, 44, 55, 79
Profiles, topographic, 65, 68, 70
Pumice, 82, 103, 117

Radiolarians, 7
Radiometric age dates, 4, 25, 44, 55, 79
Roches moutonées, 73
Roof pendants, 75
Routes of field trips, xiv, 6, 8, 10, 12, 14, 15, 20, 22, 31, 35, 40, 86, 89, 90, 105

Sand, beach, 7, 13, 29, 33
Sand dunes, 13, 34
Sandstone dikes, 39, 40
Sandstones, 4, 19, 29, 36, 76, 111, 112
Scale of geologic time, 119
Scarps, fault, 43, 103
Sea-floor spreading, 5
Sea level changes, 30
Sedimentary structures, 19, 29, 32, 35, 38, 76

Sedimentation, 5, 7, 13, 19, 29, 33, 38, 48, 69, 81
Serpentine, 7, 112
Shells, granitic, 60, 69, 71, 72
Shield volcanoes, 85, 92, 102, 105
Shorelines, 7
Siltstones, 5, 11, 16, 19, 24, 32, 37, 38, 41, 45, 76
Slates, 45
Springs, 81, 88, 92, 109
Stairway, glacial, 66, 70
Submarine landslides, 23, 38

Tar sands, 39
Terminal moraines, 78, 80
Terrace deposits, marine, 26, 28, 32, 38
Tidal flats, 7
Tide pools, 36
Tills, 66, 77
Time scale, geologic, 2, 119
Topographic evolution, 10, 62
Topographic maps, xii, 10, 12, 14, 15, 22, 31, 35, 40, 52, 90, 106
Turbidity currents, 5, 19

Unconformities, 29, 32
Uplift, 7, 13, 19, 30, 38, 43, 62
U-shaped valleys, 51, 60, 62, 66, 69
Ultrabasic rocks, 5, 18, 112

Volcanic rocks, 5, 37, 45, 80, 85, 88, 92, 95, 104, 112
V-shaped valleys, 49, 50, 62

Waterfalls, 45, 61, 65, 109
Wave-cut benches, 7, 13, 23, 25, 28, 31, 34, 38, 41, 81
Wave erosion, 7, 22, 25

## INDEX OF LOCALITIES AND SPECIFIC FEATURES

Arch Rock, 51

Bean Hollow Beach, 36
Black Butte, 116
Bloody Canyon, 81
Bolinas Lagoon, 7, 10, 12
Briceburg, 48
Bridalveil Fall, 53, 59

Bridalveil Granite, 56, 59
Brokeoff Mountain, 88, 97
Bumpass Hell, 93
Burney Falls, 109
Burney Mountain, 111

Calaveras Assemblage, 45, 49
Cascade Range, xi, 85

Castle Crags, 114
Cathedral Peak Granite, 55
Cathedral Rocks, 53
Chaos Crags, 97, 103
Chaos Jumbles, 97
Church Bowl, 55
Cinder Cone, 98, 106
Coast Ranges, xi, 1

# Index

Davenport, 38
Devastated Area, 98, 101
Devil's Slide, 24
Diamond Point, 92
Drakes Bay, 14
Drakes Beach, 14
Drakes Estero, 12

El Capitan, 53, 58
El Capitan Granite, 55
Ellery Lake, 76
El Portal, 51

Fantastic Lava Beds, 106
Fort Baker, 6
Franciscan assemblage, 2
Franciscan core complex, 2, 4, 6

Glacier Point, 55, 61, 68
Great Basin, xi, 80, 84
Great Valley, xi, 99

Half Dome, 61, 69
Half Dome Quartz Monzonite, 55
Hat Creek Rim, 104
High Cascade Range, 85
Hot Rock, 102

Indian Flat, 50

Johnson Granite Porphyry, 55

Klamath Mountains, xi, 112

Lassen Peak, 95
Lassen Volcanic National Park, 85
Leaning Tower Quartz Monzonite, 56, 59
Lee Vining Canyon, 76

Little Hot Springs Valley, 92
Little Yosemite Valley, 61

Magee Mountain, 111
Majors, 39
"Map of North America," 58
Mariposa Slates, 45
Merced Canyon, 49
Merced River, 48, 59, 63
Modoc Plateau, xi, 84, 103
Mono Craters, 81
Mono Lake, 81
Montara Beach, 26
Montara Granodiorite, 19, 25
Montara Mountain, 19, 25
Montgomery Creek Formation, 111
Moss Beach, 28
Mount Dana, 75
Mount Shasta, 98, 115

Natural Bridges Beach, 40
Nevada Fall, 58

Olema Creek, 10
Olmstead Point, 72

Pebble Beach, 36
Pescadero Beach, 34
Pigeon Point, 36
Pine Creek, 10
Point Reyes, 6, 16
Point Reyes Beach, 13
Point Reyes National Seashore, 11, 14
Point San Pedro, 19
Prospect Peak, 98
Purisima Formation, 32

Raker Peak, 102
Royal Arches, 61
Saddlebag Lake, 76

San Andreas fault, 6, 10, 18, 38
San Francisco earthquake, 10
San Gregorio Beach, 31
San Mateo Beaches State Park, 26, 31, 35
Santa Cruz, 44
Santa Cruz Range, 1, 18
Santa Lucia Range, 1, 3
Sentinel Granodiorite, 55
Shastina, 115
Sierra Nevada, xi, 43
Siesta Lake, 72
Subway Cave, 104
Sugarloaf Peak,
Sulphur Works, 85
Sur Series, 3

Table Rock Beach, 40
Taft Granite, 55
Tenaya Canyon, 61, 73
Tenaya Lake, 72
Tioga Lake, 76
Tioga Pass, 74
Tioga Road, 70
Tomales Bay, 7
Tunnel View, 59
Tuolumne Meadows, 73

Upper Merced Canyon, 48-51

Valley View, 51
Vernal Fall, 58

West Cliff Drive, 41
Western Cascade Range, 85
Western metamorphic belt, 45

Yosemite Creek View, 72
Yosemite Falls, 58
Yosemite National Park, 46, 51
Yosemite Valley, 51, 62